AF822073

CYBERMINISTRY

5 ESSENTIAL FUNCTIONS OF TECHNOLOGY IN THE MISSION OF THE CHURCH

SERIES: CYBERMINISTRY 1

Unless otherwise specified, all Scriptures taken from the Holy Bible, New International Version® NIV®, Copyright© 1973, 1978, 1984, 2011 by Biblica, Inc.™ Used by permission of Zondervan www.zondervan.com. All rights reserved worldwide. The "NIV" and the "New International Version" are trademarks registered in the United States Patent and Trademark Office by Biblica, Inc.

Sometimes, The New Revised Standard Version, Copyright© 1989, has been cited with permission by Thomas Nelson Publishers.

All references to the Bible have been taken from their digital version with permission from Bible Gateway, HarperCollins Christian Publishing, www.biblegateway.com.

CYBERMINISTRY

5 Essential Functions of Technology in the Mission of the Church

© 2023 Vladimir Lugo
vladimir@vladimirlugo.com
www.vladimirlugo.com

All Rights Reserved. All partial or complete reproduction of the content in this book is strictly forbidden unless you receive expressed written consent by the author.

Published in the United States of America
by Tepui Media, LLC, Sheridan, Wyoming

ISBN-13 Paperback:	978-0-9973369-6-2
ISBN-13 Electronic:	978-0-9973369-7-9
ISBN-13 Audio:	979-8-9884611-0-4

Library of Congress Control Number: 2023909258

Other resources by Vladimir Lugo

Cyberminister Today
Receive my articles where I explore the topic of using technology in ministry and receive notifications about new publications, conferences, and seminars right in the comfort of your email inbox. Visit www.vladimirlugo.com.

#ministerialtechnologies | #cyberministry

Follow me **@vladimirlugomt** and join the conversation in social media.

Dedication

To God,
technologist and manager
of all things
visible and invisible.

– Hebreos 11:10

There is nothing so useless as doing efficiently that which should not be done at all.

– *Peter Drucker*

TABLE OF FIGURES

TABLE OF CONTENT

PREFACE

Necessity is the mother of technology.

— HUGH OF SAINT-VICTOR, SAXON THEOLOGIAN.

In 1985, I heard the words technology and theology together for the first time. I was a freshman studying engineering at the university, and while I knew I had chosen a technical career, I had no idea of the impact that technology and theology would have on my life.

It happened when Dr. Chan, a high-ranking official from the Chinese government, visited my hometown of Caracas, Venezuela, and we invited him to our church to share his testimony. With drama and intrigue, he narrated the details of his conversion to Christianity.

As an international representative of the Communist Party, Dr. Chan dedicated his professional life to promoting Marxism, Maoist government, and atheism as the three pillars on which the ideal society is built. With dedication and effort, he climbed the highest ranks of official positions, earning the complete trust of his superiors. However, something unexpected happened when he reached the peak of his political career.

Dr. Chan received a modest invitation from a friend to join a Christian prayer group. He did not know what to expect, but he attended, nonetheless. That day, for the first time, he heard about the power of God in Christ Jesus for the salvation of humanity. He did not understand everything they told him, but he decided to continue attending the group regularly. Soon after, he became a believer.

From that moment on, the world he had constructed began to crumble beneath his feet. His philosophical framework lost its meaning. He began to sense a new fire burning in his heart and felt the need to nurture and embrace it. His worldview and interests started to change as he learned more about his newfound faith.

It was not easy for him, but over time, he concluded that it was not the absence of God but rather His intentional and merciful presence that could establish the foundations of a society that is not only better but complete.

Dr. Chan soon realized that his government position and social status were at risk. The church had been persecuted for years by the authorities, and he had become a kind of modern-day Apostle Paul, transitioning from persecutor to being on the list of the persecuted. That is why he hid his faith in the secret basement of the modern catacombs. Ah, but not for long. Dr. Chan filled himself with courage and, risking his political status and exposing himself to the death penalty, decided to publicly declare his faith. With unparalleled cunning, he requested an extended permit from the government to come to the United States to pursue doctoral studies in... THEOLOGY.

By an act of providence, and for reasons he could never explain, the Chinese government not only granted him the permission by approving his application but also sent him a letter stating that they would cover his travel expenses, room and board, and tuition costs during his studies. He read the letter in our presence, and it concluded with lines like these: "The People's Republic of China considers it of utmost importance to support a trusted government official in studying the latest global advancements in... TECHNOLOGY."

What a holy confusion it was!

He had requested permission to study theology, and they granted it, thinking he would come to study technology. This misunderstanding allowed Dr. Chan to be authorized to travel to the United States. Once the confusion was cleared, he decided to study the cause of technological progress in the Western world. During his studies, he concluded that the foundation of progress and technological development resided in the transformative message of the Gospel and its Christian work ethic. The Chinese government wanted to know the secret to technological progress, and Dr. Chan's research confirmed that the secret, fundamentally, was Christian theology. His mandate to study technology ultimately resulted in theological studies.

Theology had worked in favor of technology, and I am convinced that now technology can return the favor to theology. This is the essence of this book: Technology can work in favor of practical theology in the various ministries of the church. In other words, the church needs to explore the benefits of well-utilized technology guided by well-informed theology.

Theologian and Technologist

We, the church of today, live in an era of many challenges. The Great Commission before us reveals our need to engage in ministry as a multifaceted endeavor. The Ministry compels us to consider the use of every tool, methodology, or technique at our disposal, in any existing discipline, to effectively communicate the marvelous message we have "to the ends of the earth" (Acts 1:8).

I am a theologian by vocation and a technologist by profession. This intersection gave birth to this book: my desire to be an effective minister in our highly technologized contemporary society. The following pages are filled with personal reflections filtered through years of interactive research and professional practice. They are intended to assist ministry leaders in making the best possible use of technology with practicality and a critical eye.

I am a theolgian
by vocation
and a technologist
by professsion.

Throughout my thirty years of Christian life and ministry, I have seen how many leaders perceive technology and theology as being in philosophical and antagonistic conflict, as if they were mutually opposed. As far as I am concerned, I have not been able to validate this conflict with facts, reason, or experience. The only tension I have witnessed is more personal than metaphysical, and it has to do with my own vocational orientation and work experience – the challenge of fulfilling my calling as a bivocational minister while pursuing a career as an IT professional. This is the same tension I feel when deciding which of the following two paragraphs I want to write first.

As a theologian, I maintain a deep interest in the church and its mission to communicate the gospel or the good news of the kingdom of God, to educate, and to make disciples. In this sense, I have been a true evangelist. I have had the opportunity to express this liberating message through methods I consider both conventional (preaching, teaching, writing) and unconventional (dance, drama, music, visual arts, and computer technologies). I have used these methods while working as a youth and adult pastor, small group leader, Bible school teacher, university professor at some Christian institutions, and conference speaker.

As a technologist, I have worked in various positions promoting and implementing the strategic use of technology. Although I have not held the official title, I have been a technology evangelist. Yes, you heard it right. Even though the term may sound strange to you, "technology evangelist" is the preferred title of most technology companies in the United States for their marketing employees. It surprises me to know that the industry has borrowed this word, originating from Christian mission, to communicate the good news of their technological advancements.

I have also worked as a technical instructor, high-tech systems administrator, and IT manager for over two decades. I have had the opportunity to specialize in enterprise-wide systems that support critical applications for several universities in the country. This includes database management, Enterprise Resource Planning (ERP) systems, and multi-platform web application development. My experience has allowed me to serve as a bridge for the implementation, use, and strategic planning of technology in the institutions I have worked for.

A Mother Deserves Respect

In the early twelfth century, Hugh, a theologian, philosopher, mystic, and pedagogue from the Abbey of Saint-Victor in Paris, presented his fourfold classification of sciences based on the presupposition that all knowledge comes from God and the study of theology, the same conclusion reached by Dr. Chan.

According to Hugh of Saint-Victor, the sciences are classified as Theoretical, Practical, Mechanical, and Logical. Although the term "technology" was not in use during the Middle Ages, the practical and mechanical sciences encompassed all the developments that we would now call technological.

Hugh's writings were highly influential in the origins of Scholasticism and served to promote the acceptance of the "new scientific method in Catholic schools,"[1] in the very beginnings of universities. It is not surprising, then, that Hugh is the most cited theologian by modern technologists. In his work, *Didascalicon*, he writes, "Learn from everything; then you will see that nothing is superfluous,"[2] meaning that all present knowledge will converge at some future point. *Cyberministry* represents that future convergence for me, and I hope for you as well.

When I was young, I learned from my grandmother the saying, "Idleness is the mother of all vices." Later, when I was in high school, a friend jokingly added, "...but she is a mother, and we must respect her." Also in high school, I heard, "Necessity is the mother of invention." This phrase is attributed to Hugo of Saint Victor, although, he wrote, "Necessity is the mother of technology." In any case, this book seeks to respect that mother called necessity.

There is an enormous necessity for a theoretical and practical framework regarding technology in Christian ministry. We need an appropriate model for thinking about technology. We also need an appropriate model for planning, implementing, and utilizing technology in ministry.

Why This Book?

Cyberministry provides this framework. It simultaneously offers a critical guide and an organizational model. It has no comparison or precedent in any other work on the subject. It fills a void in Christian literature regarding the use of technology in ministry, even creating a new terminology and category of reference, as we will see later: ministerial technologies.

The book *Cyberministry* is the first volume in a series of books on ministerial technologies. Each book explores a different dimension of technology in mission, like a prism that reveals shades of colors, each one refining the light of our understanding.

I want to clarify that, although I occasionally discuss the impact that communications, transportation, or audiovisual media have on ministry, my firsthand experience comes from the field of computer and information technology. I know that this could initially be seen as a limitation in my ability to write this book, given the wide variety of other technologies. However, I consider it an advantage for two reasons.

First, we cannot deny that computing has had a tremendous impact on our culture and way of life. Computing initiated a new technological revolution of global proportions, permeating every industry and human experience. And second, there is an evident need in local churches and other ministries to keep up with the latest advancements in computing at all levels.

Whenever necessary, I have consulted various experts and authorities in other fields to ensure that the reading informs and challenges you.

Who Should Read It?

Cyberministry is aimed at Christian leaders and workers in various organizations, whether in churches, missionary and parochial organizations, educational institutions, community service foundations, or even in the job market.

Pastors

Those who seek to understand contemporary times by defining the best strategies to reach their generation with the message of salvation and lay the foundation for future generations. This includes senior pastors, administrative pastors, and pastors of various ministries or departments. This book will help them share a common vocabulary and develop clear strategies for ministry mediated by today's technology.

Denominational Directors

Those who develop policies and procedures that identify, define, and lead the representatives of their denomination. This includes regional superintendents, national directors of church centers and pastoral associations, and other apostles and pastors of pastors.

Together, they can tackle today's technological challenges to formulate a comprehensive ministerial vision.

Christian Educators

Those who lay the foundation of our worldview, instruct our children to live according to Christian values, and are committed to the theological, ministerial, and holistic formation of our leaders. These include professors in theological seminaries and universities, teachers in schools at various academic levels, from elementary to higher education, as well as Bible study instructors. They need to take the lead in guiding their students, even in matters of technology.

Leaders

Those who carry out the day-to-day practical ministry tasks. The worship leader and musicians, sound and media production team, evangelism coordinator, missionaries, counseling team, social workers, small group facilitators, administrative and managerial staff. All of them will find practical advice to be more effective in their ministry work by incorporating technology into their regular routines.

All Christians

Those who have been called to the Great Commission. The mandate of Jesus Christ remains relevant today and faces the new challenges of technological advancements. It is our task to find new ways to keep it relevant and transformative.

Cyberministry has something for every Christian, if they desire to make better use of technology in the exercise of their ministerial functions.

How to Use It?

Cyberministry is written with four dimensions in mind. Each of these four dimensions intertwines and complements each other to present a comprehensive vision of technology in the ministry of the church:

1. **A biblical and theological dimension**, so that you can distill principles and develop practices endorsed by the Word of God. This dimension will enable members of your ministry team to regain the necessary confidence to advance in fulfilling the mission.

2. **A theoretical dimension**, that will help you place ministerial technologies in the perspective of existing concepts in the information industry and create a framework for thinking about technology in mission.

3. **A practical dimension**, to help you organize technology teams in complete alignment with your mission, document existing processes for implementing improvements, and measure the results of your initiatives both qualitatively and quantitatively.

4. **A human dimension**, because we often forget the role that people play in technological matters, and as I usually say, technology is by the people and for the people. In this regard, technology aligns with the mission of the Church, which is also by the people and for the people.

Cyberministry develops the concept of ministerial technologies and explores it from the perspective of the three fundamental activities of technology in **Chapter 4**, and the five essential functions of the Christian mission from **Chapter 5** onwards. While it can serve as a reference material after reading, I recommend that you read them progressively to build upon the knowledge gained from previous chapters.

At the end of the book, you will find two appendices: the first one for assessing negative techno attitudes in your ministry, and the second one with discussion questions to engage with your team.

I invite you to contact me personally. You will find my contact information in the **About the Author** section. Join my book discussion group on Goodreads, the website dedicated to book lovers (https://goodreads.com/vladimirlugomt) and find me on social media with the unique identifier: @vladimirlugomt.

Technology is by the people and for the people.

I invite you to engage in a conversation with me on this topic. To do so, I would like to hear your opinion about the book, which you can share privately through my personal contact. If you are willing to go the extra mile, after finishing your reading, please leave a comment about the book on the Amazon page. Your honest appreciation and feedback can help me better assist you in future editions and other publications on the subject.

I hope this book helps you recognize the importance of using technology in the effective exercise of ministry while assisting you in incorporating technology into your pastoral practice. I hope you become a cyberminister, an evangelist of ministerial technology in the place where God has called you to serve. I also hope, and it is my prayer, that by the end of the book, you will have gained a better understanding of how technology can serve the purposes of theology in spreading the seed of God's knowledge.

With great expectations, yours truly,

Vladimir Lugo
Los Angeles, California
June 2023.

CYBERMINISTRY

INTRODUCTION

Technology imitates nature. 3
— ARISTOTLE, GREEK PHILOSOPHER.

I was finally sitting in the computer lab. It was the only lab at Simon Bolivar University, and at that time, the most advanced in the country. It had the first Kaypro II portable computers. Of course, back then, "portable" meant something quite different from what it means today.

The Kaypro II was the size of a large suitcase, with the keyboard on the lid that closed like a trunk and a big handle to carry it because it did not have wheels. It was so heavy that it required two people to carry it. But it was portable compared to those computers that occupied an entire floor.

The popularity of the lab made it almost impossible to reserve time alone with the computers. I was studying my first programming class. Finally, I had my chance! The day I sat down to work on a computer for the first time, my love-hate relationship with technology was born. Perhaps I should say hate first and then love.

I started off on the wrong foot. I spent many hours writing code routines and manually testing my programs until I could have my chance. But my programs were full of errors.

Whoever came up with the saying "third time is the charm" must not have had me in mind. It did not work for me. I had no luck writing a functioning program for my programming class. Since a functional program was a course requirement, I failed three times in a row.

Looking at the bright side, this gave me the advantage of studying programming for a year and a half. On my fourth attempt, Basic and Casio came to my rescue. These were not the names of my university friends, although thinking about it, they became great allies.

I remember leaving the classroom with several friends after the professor announced that I had passed. The program I wrote finally worked after so many attempts. That day I shouted, "I hate computing!" and I promised myself, "I will never touch a computer again for the rest of my life!"

But I had unknowingly fallen in love with Casio. The Casio PB-100 was the first programmable pocket computer.[4] The beauty of this computer was that it had a full keyboard and all the Microsoft Basic programming commands already predisposed on that keyboard. It also accepted an extended memory module of... 2KB, imagine that! To put it in proportion, the Casio would be 35% larger than my mobile phone, but the memory would be the size of a microscopic biological virus on the screen.

I had also fallen in love with a girl from church. A poor and neglected high school student who needed tutoring in physics and chemistry... and there, our chemistry was born... Oops, sorry, I digress. What does one thing have to do with the other? Well, when the time came to confess my love—to the girl, not Casio—I wrote a small question-and-answer program in Basic and installed it on the Casio.

I waited until she had already said goodbye. She had closed the gates protecting the apartment door. I do not know why, but that gave me a sense of security, so I handed Casio through the gates, and she started running the little piece of software I had written for her, amazed by the novelty. Who said programming language could not be used to recite a poem?

What she did not know was that the final question in my program to ask her to be my girlfriend, or in good slang, "Do you want to 'go out' with me?" had only one answer: "Yes," of course, and if she chose that option, the program would continue with its line of questions and answers in an endless loop until she had no other choice. When she willingly said yes, I was filled with excitement. I said, "Then, we will talk tomorrow." I took her hand and kissed her. Casio... the girl was behind the gates, remember?

Although many other programs failed, the program I wrote in Basic for Casio worked perfectly because that girl is now my wife of over thirty years.

Divine providence

I could never keep my university promise. As soon as I entered the job market, I had to use computer terminals connected to a mainframe. Later, I used other computers as they populated office desks: Mac II, IBM with DOS, and then *Microsoft Windows 3.1 for Workgroups.*

I worked at the first mobile phone company in Caracas, programming the Motorola "bricks" before delivering them to customers. This experience took me to Puerto Rico to work with cell phones and exposed me for the first time to designing computer networks.

When we decided to move to California for me to study at Fuller Theological Seminary, I started working as a computer instructor at a vocational school. That is when I truly decided to pursue this field. By the time I graduated with a Master of Divinity degree after three years, I had also learned enough about computing to make it a career.

I am convinced that divine providence guided me in this process. During my years of studying theology at the seminary and teaching at the technical school, the idea of uniting both disciplines formed in me:

the rich tradition of Christian doctrine and practices alongside a good relationship with contemporary technology.

Every Christian ministry needs to remain faithful to its confession of faith, traditions, and ethical positions, in short, to its mission. At the same time, it experiences tension with its traditional methods as it tries to relate to a highly technified culture full of options. This tension is healthy, but where is the balance? What will be the most appropriate and effective way to use technology in ministry?

Cyberministry not only aims to guide the conversation about ministerial technologies in the future but also seeks to answer these questions from a strategic standpoint of the mission of the church. The divine providence that guided me will also guide us in this process.

Strategic approach

In my experience, many Christian ministries rush to buy devices and software and adopt certain technological tools without a comprehensive understanding of the technology and, worse, without thoroughly considering the role that technology plays in the exercise of their mission. Many other ministries resist, either due to theological stances or because they feel overwhelmed by the reality they face.

In my opinion, the right and balanced approach is achieved by developing a strategic focus that aligns technology with the mission of the church. Strategic alignment is precisely this: you devise the commission before you commission the device.

Cyberministry offers a strategic framework by aligning the use of technology with the five essential functions of the church, and therefore, of every Christian organization: adore, announce, assimilate, assist, and associate.

Cyberministry approaches the topic from a professional perspective. The book is the result of twenty years of interest and research by a bivocational minister who serves God simultaneously in higher education and ministry. He is convinced that not only can ministry benefit from technology, but contemporary technologies can also inform today's theology, just as ancient technologies informed the theological truths of the Bible.

The bibliography consulted regularly over the past twenty years suggests that this book fills a significant gap. While technology is considered strategic for achieving goals and a tactical advantage in the competitive market in any modern organization, there is no book that addresses the topic of ministerial technologies from a strategic standpoint of mission.

Some books about science and religion or even the use of certain technologies in the church can be found, but many of these books have been written by academics rather than technology practitioners. Other authors have addressed how to use specific technology in ministry, but with a tactical focus, reducing their material to the "how" without exploring the "why" and the "for what purpose." Even in addressing the "how," they do so immediately without considering sustainable models for technology in the medium and long term.

One of the authors who has addressed related topics extensively is Quentin Schultze, with several books on the use of specific technologies in ministry. In this category, we also find John Jewell, Scott Watson, Tim Welch, and Andrew Careaga.

Some authors with a more theological focus are Brent Waters, Albert Borgmann, and Richard Rojcewicz, following Heidegger's philosophy. Lawrence Terlizzese and Derek Schuurman, Justin Wise and Craig Detweiler also fall into this category, but they do not explore the topic from the perspective of missional theology and practical pastoral ministry.

If you want to explore the strategic aspect, you must turn to secular authors such as Michael Allison, Jude Kaye, Jodi Podolosky, Terry Schmidt, and Frank Medel-Gonzalez, who address the topic of business strategy. Aubrey Malphurs, Will Mancini, and Margaret White bridge the gap to ministry strategy, but although they acknowledge the strategic role of technology in ministry, they do not develop it.

As you can see from the names of the authors mentioned, the Spanish Christian literature on the topic is very limited. I will mention three works as examples. Matias Paterlini, who is a specialist in social networks, justifies their use in ministry from a practical perspective. Juan Pinto's work is interesting as he studies papal proclamations regarding communications and evangelization in the new century with the aid of technology, although he quickly moves on to discussing how certain technologies can be used in evangelization.

We also find the Italian author Antonio Spadaro in Spanish with his excellent pioneering work, which has a more theological focus. Although Spadaro acknowledges that he is not a technology specialist, his observations are very relevant and accurate. If you have not had the opportunity to read him, I highly recommend it.

Different and innovative

Cyberministry is different and innovative. It combines a biblical theology of technology integrated into strategic ministry planning, both of which inform each other for a balanced ministerial practice.

As this topic has been explored to a limited extent in the manner presented here, it has the potential to establish itself as study material in seminaries, Bible institutes, and conferences to begin shaping the new generation of leaders from the ground up.

It is likely that our current leaders, whom we affectionately refer to as the "old guard," may feel intimidated by this topic. But even if you consider yourself part of this group, I encourage you to keep reading. There is nothing in this book that is beyond your reach, I promise you, and I am confident that its pages will help you develop your strategic vision of technology.

Cyberministry *combines a biblical theology of technology integrated into strategic ministry planning.*

Cyberministry consists of eleven chapters that progressively build the complete model. I begin by reviewing the common attitudes of the church toward technology and then move on to my practical definition of technology in general. After explaining the three fundamental activities of all technology, I will go on to define the concept of ministerial technologies and delve into its details. I then conclude with references to the various applications mentioned throughout the book.

The book contains abundant technical references and resources that can be practically utilized to guide any Christian organization or ministry in generating their own strategic plans and goals and achieving the best results in their use of the different technologies discussed here.

Due to the ever-changing nature of technology and the static nature of printed text, and of course, given the very subject of the book itself, I have decided to accompany the book with a website bearing the same name.

Make sure to share your email address with me so that you can stay updated and access document templates and technical resources for ministry by visiting the website: www.cyberministry.info.

OTHER EYES

The only true voyage of discovery... would be not to visit strange lands but to possess other eyes. [5]

– MARCEL PROUST, FRENCH NOVELIST.

My youngest daughter enjoys accompanying me on shopping trips. On one occasion, after we had finished going through the aisles of the supermarket, we approached the checkout line. She looked around the store in all directions, and I observed her. Suddenly, she asked me, "How many numbers do you think are in this store?"

Her curiosity sparked my own. I understood that she was referring to the number of digits printed in some way. Together, we began to inventory all the things that were in sight and contained digits. We concluded that the number of digits in the store must be in the order of billions. Numerals are certainly so present in our daily lives that unless we are sticking labels in the supermarket or studying mathematics or accounting, they go completely unnoticed.

Take a moment and look around you. Make a mental note of what you see. Note down those things that belong only to the realm of the printed. This book, of course, tops your list if you are reading it on paper, of course. Now broaden your view to the immediate environment. You will surely find letters, signs, advertisements, instructions, labels, illustrations, magazines, posters. In short, a whole range of categories of the printed.

Count how many objects you see within those categories. For example, as I write this section of the book with my customary morning coffee and without getting up from my seat, I see the illustration printed on the cup and the seventy-six keys on my computer keyboard, with their ten Indo-Arabic numerals, twenty-six letters of the Anglo-Roman alphabet, and other symbols and abbreviations - actually, this keyboard has seventy-seven, but the space bar has nothing printed on it.

I continued to look around and noticed the print on my T-shirt, a sticker promoting a technology company, the logo of a sports brand, and the logo of a sound equipment brand. I see two labels stuck to the lamp's cord, a small trunk with a printed design that has been applied with papier-mâché technique, five photographs, four greeting cards, and a postcard sent from Europe.

Just like the number of digits surrounding us, it is very easy to overlook printed things. They are so obvious. They are so present that we rarely stop to think about them. They are part of the landscape. We overlook their ingredients and their history. The same goes for the surrounding technology. Technology is so ubiquitous that it tends to become invisible.

The Latin alphabet, which was created around 700 BC, pulp paper, which was invented by the Chinese around 100 BC, and hydraulic mills for its mass production, which were created by European monks in the thirteenth century, are fundamental pieces of printing technology. But we don't think about them when reading a letter.

It is easy to overlook that the Indo-Arabic digits we commonly use are about 1,500 years old, printing was invented in China over a thousand years ago, and movable type was invented by Johannes Gutenberg in Germany over five hundred years ago. And of course, we omit that the direct ancestor of the keyboard is the teletype, created less than a hundred and fifty years ago, and that the keyboard and the computer are less than seventy years old.

> *Technology is so ubiquitous that it tends to become invisible.*

Come and See

The Bible tells the story of how Jesus chose His disciples. His first two followers, Andrew, and his brother Simon, were recommended by John the Baptist, who came to prepare the way of the Lord not only through his preaching but also by referring his own people to Jesus. When Jesus realized they were following Him, He extended a simple invitation: "Come… and you will see" (John 1:39).

The story says that they spent the night with Him, and the next day He found Philip and invited him to follow Him. Philip, in turn, found his brother Nathanael and invited him in the same way that Jesus had invited Andrew and Simon, "Come and see" (John 1:46). The perspective on life for these men was never the same again.

There are different levels of seeing. We can look at printed things as decoration. They are there, entertaining our eyes and adding color and splendor to the image. We can also see printed things to understand the message they want to communicate. Words of congratulations, a memory from the past, or the promotion of a product or event are typical messages of the printed medium that, without the intention of truly seeing, become visual noise.

We can also observe more deeply and realize that the world of the printed relies on the many technologies that support it. We could expand the technological list to include pigments and inks, paper, plastic, machinery, photography, transportation systems, and postal services, among other things.

I am confident that the simple visual exercise above can change our capacity to observe. This can be applied to other things as well. For example, we can look at an open Bible on the entrance of a house with Psalm 91, but we can also read the psalm or engage in a deep exegetical study of the text. Each of these interactions with the Bible represents a different level of seeing.

I invite you to see with different eyes: Come and see! Come to see with greater depth. Come to see with the purpose of understanding and acting intelligently. Let us together engage in an exegetical study of technology and its use in Christian ministry.

Visual Inventory

For now, let me expand on the categories of what you see and continue the visual inventory of technology. It is not just the printed materials that surround you, of course. You are dressed, I hope. The various garments with all their details are technology. You might be sitting, perhaps. The furniture and all its accessories also qualify as technology. It is even possible that you are enjoying a snack. The production of those foods, their transportation, and packaging involve intricate engineering processes.

Do you need me to continue with the list? Alright. The floor, the ceiling, and the walls of the building where you are constructed according to standards that must adhere to city codes. You are most likely in one of its rooms or halls, preferably close to a bathroom. In the bathroom, there are plumbing systems and potable water; perhaps some personal hygiene items, which are also technology.

The building, by the way, has a physical mailing address. By this, I mean that outside there are sidewalks, lighting, streets, pavement, and planters. That is what you see. And what about what you do not see? You know that the pipes, potable water, electrical wiring, and gas lines run through underground channels throughout the city. In fact, the largest technological feat ever invented by humans is the city, with all its intricate networks and interactions.

The largest technological feat ever invented by humans is the city.

You can continue the list according to your own interests. I am sure that at this point in the reading, you have already realized it. Technology distinguishes human experience. I have not presented my definition of what technology is yet, you must wait for **Chapter 3** for that, but you will agree with me that we intuitively recognize technology and engage with it daily.

Characteristics of Technology

With the visual inventory you just did, you have taken a journey from the words of the book to the big city. I did this intentionally to broaden your perspective on the things that fall under this category we call technology. Through this exercise, it becomes easier to notice some general characteristics of technology.

Industry

All technology is a product of human ingenuity, creativity, and industry. Industry is a human skill capable of taking a medium, subjecting it to a series of operations, and transforming it into another form for specific purposes. This is how iron, abundant in its natural form, becomes the blade of a knife for cutting tomatoes, or how the letters of the alphabet become language for communication or programming for computing.

This creative capacity comes from God. The second word in the Hebrew Bible is the word we translate as "created." We find it in

Genesis 1:1: "In the beginning, God created the heavens and the earth." It then goes on to detail the various animate and inanimate things that God created. It is my opinion that this creative capacity is one of the indicators that we were created in the image of God (Genesis 1:27).

Industry allows for variety. Other creative endeavors will create other technologies. Hence the variety of knives, languages, or software. This technological diversity is then related to human diversity and its creative capacity. You can see this by observing your surroundings.

Ubiquity

Technology permeates the human landscape in all cultures and all places around the globe. Certainly, not all cultures have reached the same level of technological development, but all have developed a technological environment that identifies them. That is one of the reasons why the discovery of pottery in archaeological excavations is so important. It provides evidence of human presence. Its characteristics distinguish one settlement from another.

One of the first mentions of technology in the Bible appears in Genesis 4:20-22. It mentions Jabal as the ancestor of tent-dwellers and livestock breeders, Jubal as the ancestor of those who play musical instruments, and Tubal-Cain as the ancestor of blacksmiths and craftsmen of bronze and iron tools. What does this mean? It means that technology has been with us since prehistoric times.

The Merriam-Webster's Dictionary relates the word "ubiquity" to omnipresence, one of God's attributes, but it also relates it to human beings in their desire to be present everywhere and in constant motion simultaneously. [6]

In this sense, technology is also a human product: it appears everywhere and remains in constant motion and change. Language is everywhere and constantly changes. Chairs are found everywhere and are redesigned all the time. You can see this if you observe closely.

Lineage

No technology exists in a vacuum. We just saw that the Bible mentions the ancestors of builders, livestock breeders, musicians, blacksmiths, and instrument makers. This is because every technology has a lineage, a descent. You could construct a genealogical tree for each piece of technology.

In our visual exercise, I mentioned some of this regarding the computer keyboard. In biblical terms, the teletype begot the typewriter, the typewriter begot the keyboard, and the keyboard begot the virtual laser projection keyboard. However, the ancestor of all of them is the alphabet. The same can be done with other technologies.

To see the extent of technological lineage, we can go back to the time of the ancient Roman Empire. In those days, war chariots were pulled by two horses, so the distance between the wheels was determined by the width of the yoke. The Royal Road and other paved imperial roads were built to the precise size for these chariots to travel.

After several centuries, when coal mines began to be exploited in England, horses transported coal in wagons on rails. And when trams began to proliferate, they used the old roads of the same width to lay their tracks. In the 19th century, British engineers migrated to the United States and brought with them the idea of the railway. Since they had the same tools brought from the motherland, they built an extensive railway system throughout the country, following the same distance between the rails. And so, when the space shuttle was assembled in Florida with parts manufactured in several other states, the fuel tanks from Utah had to fit through the tunnels of the railroad tracks that would transport them.[7]

You could construct a genealogical tree for each piece of technology.

This is how the diameter of the fuel tanks of the most sophisticated transportation system created by humans can be traced back to the most rudimentary horse-drawn transportation system. This track gauge is calculated at four feet eight and a half inches (4' 8½"), equivalent to 143.50 centimeters, and is known as the Stephenson's Gauge, in honor of the father of the railway.

Practicality

Every technology has practical purposes. If a technology fails to fulfill the purpose for which it was created, if it ceases to meet the practical need it intends to satisfy, if it no longer offers real solutions to complex problems, it recombines, transforms, and recreates itself to fulfill its practical objectives.

One attitude I often encounter when discussing these topics in classes and conferences is a rush towards practicality. That is why much of the literature focuses on the "how" before the "why" and "what for," as I mentioned in the book's introduction. Ultimately, the impact of technology is determined by its practical application.

"For it is God who works in you to will and to act in order to fulfill his good purpose" (Philippians 2:13). Our faith must be manifested in works. This is why we speak of the practice of our faith.

Technology connects our way of thinking with our way of acting. The intrinsic practicality of technology must be connected to our vision and our way of thinking about it. We are not exempt from the responsibility of philosophical and theological reflection, nor from strategic planning to transform, with that practical quality, the environment around us. I will talk more about the practicality of technology in the subsequent chapters.

Complexity

As we will see in **Chapter 3,** technology is a complex ecosystem of interactions and mutual support. This means that one technological system supports another technological system, enabling perpetuity, growth, and innovation in both.

Kevin Kelly recognizes technological complexity with the following words: "Technological systems gain their own momentum and become so complex and self-aggregating that they form reciprocal environments for other technologies." [8]

For example, the road system relies on public lighting, which benefits from the electrical system. Cars complete this transportation system, but not without first filling their fuel tanks, which come from the oil exploitation and refining system. The coordination of all these systems makes it possible for us to move from one place to another, transporting goods produced by agriculture and food engineering.

I am sure that if you exercise your capacity to see, you can create your own complex connections of technological systems familiar to your context. Well, take a few minutes and write them down. Why do I want you to do this? Because when we start discussing the topic of ministerial technologies in **Chapter 5**, you need to have developed a perception of the complexity of technology in service of the mission, which will assist you in your ministry.

Transformation

Because technology is a product of transformation - industrial, in the sense defined earlier – technology also produces social transformation itself. In other words, a technology can be used to transform another technology, but it also could transform an individual or a social organism.

The transformative capacity of a technology is very difficult to measure in concrete terms, and its effects are generally visible in generations after those who live and experience social change. This does not mean that it is impossible to measure, but rather that we must be very intentional about it.

Furthermore, technology can transform through its ability to mature. Throughout this writing, you will find references to the maturity of technology. The book *Cybermaturity* in this series delves into this topic and provides assessment tools to measure the transformative power of technology.

Technology can transform through its ability to mature.

Take a moment to look around you again. Take a piece of technology and answer this question: How does this technology help me be more efficient in my work, improve my quality of life, or allow me to communicate better? How was it before I had it? What is the difference? I am sure you will find answers to these questions.

Risks

Every technology has its advantages and disadvantages, its benefits and risks, its blessings, and drawbacks. The hot coffee cup can spill and break, burn me, and cut my feet. The car I drive can cause an accident due to a mechanical failure. The same hammer that builds a house can also destroy it.

When considering a technology plan in the ministry of the church, we need to be aware of the limitations of what technology can and cannot do, not only from a pragmatic standpoint but also from a psychological and theological perspective, that is, human behavior and its need for salvation.

I have classified the risks of technology into five categories in the article: *5 Evils Inherent to Technology we are Called to Heal.* These five categories are all related to the loss of freedom and technolatry:

1. *Technoabuses*: It refers to the misuse of technology for sinful and perverse purposes.

2. *Technoalienation*: Related to social isolation when we submit ourselves to the evil dominion of technology or when it mediates our interactions.

3. *Technoaddictions*: Different ailments caused by personal enslavement and the surrendering of free will to technology.

4. *Technoafflictions*: The ailments related to physical or mental health problems directly associated with repetitive use of certain technologies.

5. *Technoannihilation*: The possibility, for the first time in human history, of causing our own mass annihilation using certain technologies. [9]

The idea behind listing these harms is for the church to become aware of its ministerial responsibility to address and alleviate them to bring healing to humanity. This can be done by creating innovative technologies as a counterforce—for example, to rescue individuals from addiction and heal their condition—or by working with authorities to regulate the harmful use of existing technology. However, in many instances, the church has disengaged from the technological pulse and left decisions in the hands of the secular world. This is a regrettable fact.

I have guided you to look outward, and now I want to take you inward, to engage in introspection. I invite you to examine some of the common attitudes we have towards technology. By doing so, we can evaluate our own attitudes as well as those of your ministry team.

TECHNO ATTITUDES

But Lord, how am I going to appear on television now after I criticized that devil's box so much?

– YIYE AVILA, PUERTO RICAN EVANGELIST.

We demonize technology for no other reason except for the fact that we are getting old.

– GARY VAYNERCHUK, AMERICAN ENTREPRENEUR.

I consider myself part of the first generation of television. I do not remember a single moment during my childhood when I did not have a television available to watch my favorite shows. Surely some of you remember Speed Racer, Batman, The Three Stooges, Ultraman, Abbott and Costello, and the famous Mel Brooks movies. My favorites? Dick Tracy, Inspector Gadget, and Pinky and the Brain. Many of these cartoons and characters became my heroes and childhood companions.

During my adolescence, after having a personal encounter with God and attending church for some time, my cousin recruited me as a volunteer for Yiye Avila's ministry. The job involved pasting promotional posters for his miracle campaign. After brief instructions, we ended up with a bundle of those posters and several containers of glue. We worked two consecutive nights until dawn, joyfully serving God in the streets of the city center. We plastered hundreds of posters, and with each one, my anticipation grew about what would happen at Yiye Avila's campaign in the bullring of Nuevo Circo, to the point that I convinced my dad to accompany me. And he did.

What a disappointment! The preacher spent most of his teaching speaking against television or, as he baptized it, "the devil's box." I think that day was my first conflict with faith. "Poor Scooby-Doo! What did he do? " I thought, "who will come to his rescue now?"

At that time, it took a lot of effort for me to understand this radical stance, although I never fully accepted it. What do some Christians have against technology? Where does this attitude come from?

Thank God, with time, His mercy reached Yiye Avila, and he not only repented of his earlier teachings but became a pioneering radio and television evangelist in Latin America. The television channel he founded, *La Cadena del Milagro*, has grown significantly over the years with regular programming, satellite and internet broadcasts, and expansion into various print and audiovisual media. In his maturity, he discovered the value of television as a means of spreading the Gospel.

Technophobe or Technophile

The tension of the conflict described and its resolution highlight two attitudes that can be held regarding technology. In fact, these two attitudes have been a topic of conversation among philosophers and practitioners for many years, especially since the days of the industrial revolution. [10]

The technophobe is a person who is afraid or feels apprehension towards technology, and consequently, holds negative attitudes towards it. The technophile, on the other hand, is a person who loves technology and tends to attribute it with qualities that can even border on the ridiculous or exaggerated.

Elena Sierra from the University of Alicante, who has explored these attitudes towards technology, explains that:

> ...technophiles argue that the Information Society democratizes information and power, while also enabling greater rationality in decision-making and freeing individuals from routine jobs. On the other hand, technophobes believe that the abuse of computer systems leads to undesirable interference in human privacy and freedom, promoting submission to technocrats and increasing unemployment. [11]

In other words, for the ones in favor of technology, allowing information to run freely through technology empowers the individual to think and decide, and frees his or her time to engage in more productive tasks. For the ones against it, technology enslaves the individual to the techno kings and alienates us, making us more prone to worries and less creative.

Kevin Kelly also writes on this respect:

> When technology appeared among our age-old routines, it was set outside ourselves and treated like an infection. People embraced its products, but guiltily. It would have been ludicrous a century ago to think of technology as ordained. It was a suspect force. When two world wars unleashed the full killing power of this inventiveness, it cemented the reputation of technology as a beguiling satan. [12]

It is well known that mass production through technical means has left many workers unemployed, leading to resentment towards new technologies to the point that many social thinkers have expressed their doubts about their benefits. On the other hand, as a defensive reaction, others have taken on the task of extolling the virtues of technology and accepting its utopian premises without a critical mindset.

I find that these two attitudes, the technophobic opposition to everything related to technology, and the technophile's infatuation with all its advancements and promises, are like two extremes of a pendulum – to borrow an illustration from technology – swinging back and forth in search of balance.

A proper attitude towards technology in ministry would lie somewhere in the middle. The middle point is defined in terms of the mission. The role of technology in your ministry can only find balance within the mission. Exercising your mission is what will make you more effective with technology.

> *Exercising your mission is what will make you more effective with technology.*

During my Christian ethics classes at the seminary, the professor led us in a conversation about the church and culture, following the classic book by Richard Niebuhr, *Christ and Culture.* [13]

We ended up discussing the Amish as a Christian group that lived against culture. In my participation, I insisted that it was not possible. The Amish did not live against culture but rather lived by prolonging a culture frozen in the early 18th century.

The apparent technological resistance of this Christian group is not true, as they still rely on technology from previous eras even for their own subsistence. I use 'apparent' with intention, as their critical attitude and selectivity towards contemporary technology have sparked their creativity and guided them to develop alternative technologies. [14]

In fact, complete technological resistance is not only impossible but also unnatural. In nature, there are many examples of other forms of life that make technological adaptations to their environment. Just consider the ant with its elaborate labyrinth of tunnels and its storage system, the bee with the perfect hexagonal structure of its honeycomb cells, which, by the way, gives identity to cellular mobile technology with its concept of cells, or the mole with its powerful dams that serve as both a burrow and a trap for obtaining food.

The difference between these rudimentary technologies found in nature is that their creators do not have the critical capacity to challenge their models, make improvements, implement significant changes, or abandon them in favor of more efficient ones. This is the critical capacity that we do possess and are obligated to exercise considering the Christian mission and ethics.

A critical attitude will protect us from excesses and allow us to adopt a pragmatic, non-idealized stance towards technology.

No one can live devoid of some form of technology, whether ancient, modern, or alternative.

Those who supposedly advocate against technology, without realizing it, cling to forms of previous technologies. As we will see in the next chapter, technology is part of culture, and culture is part of the reality of human beings, so it is impossible to be without culture and therefore without technology. No one can live devoid of some form of technology, whether ancient, modern, or alternative.

The Bronze Serpent

When I started reading the Bible on my mobile phone over ten years ago – a practice that has now become common and convenient – a lady from the church told me, "You should read the Bible as the pastor said instead of playing on your phone."

At the end of the service, I showed her the passage of the Bible I was reading on my phone, with my notes attached to the verse, and then, to show off, I confess, I showed her the same passage in several versions in Spanish, English, interlinear, and even in Greek. In the end, she surprised me by saying, "I prefer my printed Bible." I respect preferences. What many church members fail to recognize is that the book of the Bible is a piece of technology that has been around for about five hundred years, and their apparent aversion to technology is nothing more than an attachment to previous technologies.

In the Bible, there is a well-known passage that has always caught my attention. It is the passage of the bronze serpent in Numbers 21.

In the story, the people of Israel began to complain against Moses and God, so God sent venomous snakes among them, and many of them died. Perhaps complaining against the new things of God is a deadly sin. They approached Moses and pleaded with him to intercede for them and ask God for a way out.

Instead of driving away the snakes from the surroundings, God asked them to create a piece of technology – I know, I have not defined what technology is yet, which I will do in the next chapter – but for now, let us say that this "technology" was a bronze sculpture on the top of a pole.

The pole was not erected following the codes of the electric company or the fire department, but they had to make certain calculations to ensure that the serpent could be seen from anywhere near the camp.

The tribes of Israel, except for the Levites, were supposed to camp about 900 meters away from the tabernacle (Joshua 3:4) and organize themselves around it in groups of three on each side. If we consider that the camp consisted of an estimated population of three million people, [15] it is possible that there was up to eight kilometers between the two ends of the camp (Numbers 33:48-49). This means that the serpent on the pole should be visible from a minimum of four kilometers away, assuming that the surface of the entire camp was flat. The pole should have been several meters high and have a substantial diameter at its base to be seen by the average Israelite on the camp's perimeter. For reference, look at the size of a pole that holds a highway beacon or a flag at a gas station. Literally, a work of engineering!

If you analyze the bronze serpent with the "other eyes" from **Chapter 1**, you will notice that it embodies the seven characteristics from the list.

1. *Industry*: It is the product of the ingenuity and technical expertise of the people of Israel.
2. *Ubiquity*: It could be seen from anywhere in the camp and accompanied them wherever they went.
3. *Lineage*: Their ancestors can be found in the craftsmanship of sculptures, ironworking, and the knowledge they had gained as builders for the pharaoh.
4. *Practicality*: They would look up to it and be healed from the snake venom, wouldn't they?
5. *Complexity*: As I explained, the fact that it had to be visible from every part of the camp alludes to the complexity of the work.
6. *Transformation*: The individual was transformed, the families whose members survived were transformed, and the camp was transformed.
7. *Risks*: Some people may have seen the bronze serpent as an amulet of good luck or even believed it had divine powers.

In fact, the latter happened. The rest of the story of the bronze serpent reveals the people of Israel worshiping it to the point that King Hezekiah had to destroy it (2 Kings 18:4).

The technophobes completely ignored the bronze serpent and died in the wilderness. The technophiles fell in love with the bronze serpent and turned it into an idol, displeasing God and allowing it to control their lives.

Israel had clung to the outdated technology of the wilderness, thinking that it had salvific qualities apart from God. We cannot do this. Nor can we ignore technology and miss out on the opportunities it offers us to preserve life through the message of salvation, to remind us that God is with us, and to remain relevant in the context where we do ministry.

Technology is a tool to be used, not an idol to be worshiped.

Therefore, between technophobia and technophilia, there is a range of attitudes that are worth examining from the perspective of the mission of Christian ministry.

In pursuit of balance, I recommend adopting a utilitarian stance towards technology. In other words, technology is a tool to be used, not an idol to be worshiped. When technology ceases to fulfill its function, it can be replaced without remorse and without unhealthy attachment.

Common Attitudes

Let us now list some of the most common attitudes to cultivate a proper technological awareness. Each of these poisonous attitudes has an antidote that will help you avoid the effects of snakebite, so that your technological initiatives do not perish, and you can conduct a balanced evaluation focused on the mission.

1. Ignorance

The Greek word that means knowledge, science, or understanding is the word *gnosis*. In Greek culture, it was associated with connecting theory with practice, like the Hebrew notion of knowledge that we discussed under practicality in **Chapter 1**. Many times, we do not reach this level of technological knowledge due to a lack of expertise in the field. Instead of turning a blind eye, we must make the decision to learn.

Greek	Root	English	Meaning	Example
gnosis	i + gnosis	Ignorance	I do not know what I do not know	
	reco + gnosis	Recognition	I know what I do not know	"I do not speak Greek"
	co + gnosis	Knowledge	I know what I know	"I speak English"
	pro + gnosis	Prognosis	I did not know I knew	"I did not know I could speak Greek"

Figure 1: Knowledge Levels

Figure 1 shows the four levels of knowledge. Notice the Greek root of the words in English.

Ignorance is the most basic level. The antidote to ignorance is education and training. To overcome ignorance, you must start by making a list of what you know you do not know. You must answer the questions: How do I learn about this technology? How do I train myself?

In technology matters, the ministry often ignores not only the technical details of using and applying certain technologies but also what it is, what it consists of, how to maintain it, how much it costs in its entirety, and how to maximize its potential to receive the greatest benefit possible from the investment.

For example, perhaps you do not know how to develop a website, but you already know that you do not know and that you could learn it if you wanted to. Or maybe you decide to remain ignorant about it and surround yourself with people who already have that knowledge to do it for you.

This book aims to take you out of ignorance in many areas so that you learn to identify the things you do not know and make more informed decisions. I recommend paying special attention to **Chapter 3**, **Chapter 6** with extensive references to the Bible, and **Chapter 7**, which focuses on learning.

2. Rejection

As we saw in the previous section, it is not possible to reject all technology. However, this is the most common attitude I find among Christians and their leaders.

With rejection, one falls into the error of opposing certain technologies - and even certain scientific principles - without reason, only to later admit that we were wrong and adjust. This was the case with Yiye Avila.

It is not possible to reject all technology.

Part of the problem is that by then we would have missed out on good opportunities, as well as the valuable time that could have been spent at least evaluating and, even better, planning clear strategies.

The antidote to rejection is strategic planning. We need to answer the question: How do I strategically plan for this technology? To find the answer, you can study **Chapter 4** and **Chapter 5**.

3. Blindness

I also like to refer to this as blind adoption. That is, accepting any technology simply because it is trendy without weighing it based on its mission value and missionary quality, that is, its ability to strengthen the mission and its mobilization potential. In a sense, it is the opposite of rejection, it is the acceptance of everything for any reason and for no reason at all.

By considering the mission of the church, we can make better decisions regarding technology. By developing a clear vision of where you want to lead your ministry, you can deduce the details and plan how certain technology can assist you.

The antidote to blindness is to have our eyes opened to understand how new technologies can help us achieve our vision and how they spread in the environment where we minister. It comes by answering the question: How can this technology help me achieve my vision? You can delve deeper into this topic in **Chapter 4** and **Chapter 5**.

4. Traditionalism

Traditionalism involves an excessive attachment to outdated technological models that no longer have a contemporary impact. It should not be confused with tradition. Tradition is good, traditionalism is not. Traditionalism is a negative attitude.

The tradition of faith, biblical values, the essence of the Christian life—in other words, foundational aspects—are things we are obligated to defend wholeheartedly. Traditionalism regarding forms and methods can be subject to scrutiny and adjustment to new realities.

Just as cultural development is dynamic and its study is essential for the execution of the church's mission, understanding technology as part of culture is crucial to staying relevant in a changing world.

The antidote to traditionalism is openness to the new in light of the mission, which is the main theme underlying **Chapters 6** to **10**.

5. Delays

Between complete rejection and blind adoption lies the attitude of delay. This attitude can be seen as more balanced if you consider the technological life cycle and the benefits of deferring the process. For example, delaying implementation can allow for better planning, designing a more suitable solution, or obtaining better prices when making purchases.

I address the topic of technology life cycle in more detail in my book *Cybermaturity*. In it, I present the only existing technology maturity model for Christian ministry, consisting of seven steps ranging from initiation to innovation. Taking the time to create a solid technology maturity model for your ministry is a controlled delay that can yield significant dividends in the medium and long terms.

Delays caused by politics, disorder, indecision, or any of the other negative attitudes in this list are unnecessary delays and indicate an attitude that needs correction.

The antidote to delay is to outline clear purposes and achievable goals anchored in time, if we ensure that the delay is not due to ignorance, paralysis, or unresolved moral considerations. It involves answering the question: How do I define clear purposes and achievable goals? In **Chapters 6** to **10**, I provide several tips for project implementation and management that will assist you in this regard.

6. Indiscipline

Indiscipline lowers the standards of quality, causing people to lose interest and gravitate towards other options not provided by your ministry or provided in a limited, mediocre, or sporadic manner.

Indiscipline leads to a lack of consistency. A clear example is seen when we are inconsistent in delivering content, whether it is for the website or teaching materials. Indiscipline can also cause interruptions in the technology-mediated services that the organization may offer.

Indiscipline leads to a lack of consistency.

Although it may be obvious, I must say that the antidote to this poison is discipline. Discipline is the key to success with any technology as it keeps it available and up to date. It helps us answer the questions: How can I improve standards and service quality? What practices enable regular and controlled maintenance?

Chapters 6 to **10** and the resources available on the website (www.cyberministry.info) provide various recommendations that will help you grow in this area, both personally and corporately.

7. Disorganization

Indiscipline and disorganization go hand in hand, but they are not the same thing. Having discipline means following the specific field of study of an art, faculty, or science. [16] Indiscipline, then, would consist of not following – whether due to ignorance or some other reason – the principles that govern the use of technology. This book addresses the principles of this discipline.

Order refers to the sequence or relationship between things. According to this, disorder would be, for example, wanting to buy wheels before buying a car. It is worth noting that the word used in Spain for computer is "organizer." The focus is on its ability to organize, not just compute. Technology can help us organize internal affairs of ministry as an organization and external affairs of its work in the world.

Technology systems and projects follow a particular order. Becoming familiar with the concepts of process analysis and design, as discussed in **Chapter 3**, designing technology programs according to their three fundamental activities, as discussed in **Chapter 4**, and understanding the technology life cycle to organize well in each stage of this cycle are the antidote to disorder. We replace this attitude by answering the question: How do I organize each stage of the process and life cycle? What controls and measures do I adopt to maintain order in the medium and long term?

8. Paralysis

There is a very common phrase in my profession: "analysis paralysis," which means getting caught up in considerations without taking action. Consideration without action is dead ("Faith by itself, if it is not accompanied by action, is dead," James 2:17). Again, remember what we discussed in **Chapter 1** about the importance of connecting what we believe with what we do.

A paralysis attitude is often a response to other underlying attitudes. Paralysis by intimidation is common. This attitude is, in my opinion, more understandable, as the world of technology is so vast that it can be overwhelming to know where to start and which steps to take. Aristotle wrote, "the principle seems to be more than half the whole."[17] So I encourage you to take the antidote against paralysis and start; answer the question: How do I begin working with this technology?

In my book *Cybermaturity*, you will find a model for designing an initial action plan and growing with technology.

9. Immorality

During the dot-com bubble era, I was working for a company that offered online advertising. The company was considering becoming a provider of online pornography. This was unacceptable for us as employees. Several of us stood united and presented a collective resignation. Within a few months, the company decided to shut down its operations. We never really knew what happened, but it is likely that they continued with their plans under a different name.

These days, there are many alternatives to consider when the corporate principles of a provider are contrary to Christian principles, or when the impact of certain technology goes against our faith and values. If alternatives do not exist, perhaps we are facing a need that God is calling us to fill as a ministry and an opportunity to create something new.

As I mentioned before, every technology carries risks, but I still believe that every technology has the potential to be used for ministry purposes, as long as we are sensitive to its ethical and moral implications.

The antidote for this attitude is the moral evaluation that must be done for each specific technology. It is about answering the question: How do I evaluate the moral issues associated with this technology? In **Appendix A** and the online resources, you will find templates that will help you evaluate your own values and the technology in terms of morality, among other things.

10. Infobesity

Infobesity, [18] sometimes also called infotoxicity, information overwhelm or information overload, refers to the state of being bombarded with such a high volume of information that it ends up distracting and incapacitating us in decision-making. [19]

A classic example of this is our postal mail. Every day, our mailbox gets filled with restaurant flyers we never visit, real estate agent brochures we do not need, business cards from the neighborhood gardener even though we do not have a yard, and supermarket coupons for food we do not eat, among other things. We need to sort through and eliminate a large portion of the information we receive, lest we lose that immigration letter we have been waiting for amidst the pile of papers.

And let us not forget about our email. I have noticed that many Christian institutions collect a large amount of information that becomes a burden as it is not used for anything productive. A minister or ministry should not blame technology for this problem but rather examine their own habits.

The antidote to infobesity is an infodiet, [20] which means developing good habits of collecting, selecting, using, and analyzing information. Moreover, with little effort, other technologies could assist you in reviewing the information, helping you decide what is important and what is not. It would be like calling the postal office to suspend unwanted mail for a while. To do this, I must answer the question: How do I decide which information is important and which is not?

Chapters 6 to **10** address the topic of information management in ministry.

11. Insecurity

While I write this section of the book, Apple has just admitted that its *iCloud* system has been breached by unscrupulous individuals who stole private information from several celebrities, including some compromising photos. [21] Information theft is real. It is not a matter of "if" but "when" it can happen. I understand that cyber insecurity can make us feel vulnerable and fearful of any openness to technology. The antidote is prevention and protection. We need to understand the common privacy breaches and attacks on our information, most of which can be prevented by consistently following simple principles. We need to be able to answer the question: "How do I protect and prevent information theft?"

Chapters 6 to **10** will address the topic of security as needed. The book *Cyberethics* in the series also delves into the topic in more detail.

12. Excuses

Many excuses have been used to justify erroneous positions. I have heard leaders from various organizations say things like, "An old parrot cannot learn to speak," "I just want to retire and not have to deal with it," "It is too complex for me," "What if I break something?," "There are too many options available," "We have always done it this way," and the most common of all, "It is too expensive."

Although I have addressed some of these evasions in the previous paragraphs, I want to focus on the last one: there is something for everyone. In other words, for every costly option, there are several free or low-cost alternatives available, thanks to the open-source movement, the Free Software Foundation,[22] and *Software as a Service* (SaaS) offerings in the cloud.

However, the fact that they are free does not mean they will not cost anything, as they will certainly require time, dedication, and collaborative effort to make a project successful.

The antidote to excuses is the analysis of benefits and cost-effectiveness. It is the result of answering the question: "How do I analyze and weigh the benefits of this technology?" **Chapter 10** delves a bit more into financial matters in the context of project management.

Attitude	Venom	Antidote	Chapters
#1	Ignorance	How do I learn this technology? How do I train myself?	3, 6, 7
#2	Rejection	How do I plan strategically with technology in perspective?	4, 5
#3	Blindness	How can this technology help me reach my vision?	4, 5
#4	Traditionalism	How can this technology help me accomplish my mission?	6-10
#5	Delays	How can I define clear technology purposes and attainable goals?	6-10
#6	Indiscipline	How do I improve the technology standards and quality of service?	6-10
#7	Disorganization	How do I organize for every stage in the technology life-cycle?	3, 4
#8	Paralysis	How do I start using this technology?	6-10
#9	Immorality	How do I assess moral issues related to this technology?	A
#10	Infobesity	How do I decide which information is important and which is not?	6-10
#11	Insecurity	How do I prevent and protect from cybercrime and data theft?	6-10
#12	Excuses	How do I analyze the pros and cons of this technology?	10

Figure 2: Attitudes Towards Technology

You can see a summary of these attitudes in **Figure 2** above, which you can use as a quick reference guide to deal with these attitudes if they are present in your ministry.

I recommend that you use the table of attitudes with their corresponding antidotes as a survey tool, asking the appropriate questions to identify the prevailing mindset and prioritize the immediate needs. You can use **Appendix A** for this purpose.

For example, if the most recurring attitudes in your context are traditionalism and immorality, you could start by defining the values of your organization, evaluating the moral issues associated with those technologies, and analyzing how the novelty you want to introduce could either benefit or hinder your mission. This is crucial: when it comes to technology, your mission should guide your decision-making, as we will explore in **Chapter 5**.

> *When it comes to technology, your mission should guide your decision-making.*

I hope this chapter has not discouraged you in your intentions, but rather helps you identify common attitudes that may exist personally or within your leadership or ministry body. By doing so, you can develop an action plan. After all, it is God who works in us to will and to act according to his good purpose.[23]

I have included this chapter on tech attitudes because, as Peter Drucker once said, "Culture eats strategy for breakfast," which means that regardless of the strategy, the culture of your ministry will prevail. It is the culture, with its everyday attitudes and habits, that will determine how technology is used or not used in your ministry. Therefore, if we start with an evaluation of the organizational culture regarding technology, we can address the issue and create space for a clear ministerial strategy, as we will see in **Chapter 4**.

But before we proceed, let us define the meaning of technology.

TECHNOLOGY

Technology is a dynamic, holistic system. It is not mere hardware; rather, it is more akin to an organism. [24]

— KEVIN KELLY, *WIRED* MAGAZINE FOUNDER.

The first rule of intelligent tinkering is to save all the parts. [25]

- PAUL EHRLICH, GERMAN SCIENTIST.

I was teaching computer classes at a vocational school for adult education. To be honest, I had the feeling that many of my early students were attending school more out of obligation than a personal desire for self-improvement. In fact, many of them were receiving the classes for free through a state-funded program that aimed to help them re-enter the workforce. During those days, I was teaching them how to use the productivity tools that would become the standard in every company, *Microsoft Office.*

I remember I was teaching some advanced features of the word processor Microsoft Word when one of my habitual latecomer students interrupted the class. This time, I noticed something a bit different but could not immediately figure out what it was.

I asked him to please come in and take his seat in front of the assigned computer. Mumbling incoherently, he walked down the aisle and sat down with his arms crossed. He closed his eyes as if he were asleep. The rest of my students were surprised, but without dwelling on the matter, I resumed the class and continued. After a few minutes, he suddenly stood up and shouted:

— "After all this, what's next?"

— "What do you mean?" I returned the question, with the rest of the class looking puzzled.

— "Suppose I attend all the classes, pass all the exams, and complete all the assignments. What comes next? What am I going to do with that?" By this point, we all realized he was drunk. Patiently, I explained that he was acquiring skills that would help him soon, but he persisted:

— "Yes, but what else?" I tried to explain, but he kept interrupting me whenever I tried to speak. Then I asked him to please sit down and listen to me. I told him that these classes would help him find employment... — "and what else?"— he continued asking.

— "Then you can take the official exams to get certified." In those days, the corresponding certification was well recognized in the industry, so I explained its benefits to him. The certification was called *Microsoft Office User Specialist*, or MOUS for short. With great patience, I spelled out the acronym: "M," "O," "U," "S," while writing it in big letters on the board. However, MOUS sounds like the word "mouse" in English. And that is when I made my mistake. I said to him:

— "If you get certified, you'll become a MOUS". He angrily stood up, and to the surprise of the rest of the class, he started shouting, "I don't want to be a MOUS (mouse), I want to be a LION, like you! I want to be a lion! Roar!"

When I recently shared this story with a friend, he made me realize that Loyola Marymount University, the university for which I was working at the time, had a lion as its mascot. I did not know that the words of this drunken man would become a prophecy. I do not know if he became a lion, but I did.

A Matter of Terminology

During my years of experience as a technology instructor, I have told my students that learning terminology is at least half the work. And there is certainly a lot of terminology in this field!

Millions of objects, tens of thousands of programs, thousands of functions, hundreds of programming commands - the number of words that technology, in general, adds to our dictionary every year is considerable. [26] Many words never even make it into the dictionary but remain part of popular culture or technical documentation in the hands of a small number of specialists.

Knowing the language is essential. At the very least, it helps us clarify any confusion and facilitate communication.

I often like to ask people, "What is technology?" I have noticed that many of us refer to technology, but very few fully understand its meaning. I am surprised by the kinds of responses I receive, even from those who work with me in the field of computer science.

I must confess that for years, I too worked with information technologies without stopping to think about technology as a concept, let alone trying to define it. Some people rush to answer my question and raise their cell phones. Others mumble phrases about what technology has done for them, or the impact it has had on society, or how their three-year-old daughter finds it so easy to use. But they do not quite reach a definition.

Most people think that, as Alan Key suggested: "Technology is everything that wasn't around when you were born," [27] unaware of previous technological advancements that are so prevalent in everyday life that they go unnoticed. We addressed this in **Chapter 1**. Many individuals I have interacted with in churches and Christian ministries believe that technology is limited to the devices we use, but it is much more than that.

Several years ago, a pastor from a church I attended, which I will refer to as Project "R," told me, "We're getting into that technology stuff: the church is buying a projector." During the conversation, I learned that the projector installation and screen had a budget of twenty thousand dollars, including the construction of reinforced supports to hang them from the ceiling. After a few days, there it was, a projector the size of a suitcase looming over the heads of the congregation.

After several weeks, I asked, "Why are we not yet using the projector?" and they replied, "The image is not visible." The initial plans had not considered the lighting conditions of the place, which had abundant natural light, to choose the most suitable equipment. Shortly after, I noticed that they had replaced it with a larger one. Several more weeks passed, and I asked again, "Why are we not yet using the projector?" to which they responded, "No one knows how to use it."

In the mind of this pastor, "getting into the technology stuff" was synonymous with purchasing a device. But if anything is evident in this story, it is that technology goes much further. What the pastor needed was an accurate definition of technology that would inform him of its foundations and guide him in practical matters. It was a matter of terms.

A coherent definition of technology will enable you to have a proper relationship with it. The opposite is also true. If you do not define technology correctly, you will not be able to harness its benefits effectively.

Holistic and Integrated

Now, where do we begin to define a term that has become part of everyday life? We could start by quoting, for example, the definition from the dictionary. Merriam-Webster defines technology as the "manner of accomplishing a task especially using technical processes, methods, or knowledge."[28]

This definition assumes that technology requires knowledge, whether that knowledge derives from science, as some people may think. In fact, I recall as an engineering student debating the value of technology in the university hallways with those studying "pure" science. They proudly proclaimed that without them, we, engineers, and architects would not exist. In other words, without physicists, mathematicians, and chemists, there would be no engineers or architects.

However, science is posterior to technology. Technology precedes science in its most rudimentary aspects in the development of life. Technology is more natural and even more spontaneous. As I mentioned before, we can observe other species incapable of science but capable of technology. Spiders build their webs, birds build their nests, chimpanzees use sticks as tools, and grasshoppers use big leaves as amplifiers, to name a few examples.

On the other hand, it could also be argued that science, with its methods, research, and documentation, is itself technology.

Another way to define technology considers its etymology. The word derives from the Greek *techné*, which literally means art, craftsmanship, or skill, and *logia*, which denotes discourse, study, or treatise on something. In this sense, technology is the study of art.

This translation also does not help us much since what we understand today as technology is different from what we understand as art. However, it contains the idea that technology is the product of human design and ingenuity, and it presupposes that art, in all its forms, depends on the corresponding technology for its expression. For example, painting depends on knowledge of pigments, solvents, and materials, while music depends on the creation of instruments and acoustics.

Technology is the product of human design and ingenuity.

Aristotle was the first person to combine these two terms, who mentioned them five times in his essay *Rhetoric*, [29] referring to the use of the word *techné* as art and *technologéo* as the use of rules and techniques of persuasion. So, for Aristotle, the art of speaking to convince someone of one's arguments was technology.

The term technology itself was ignored for several centuries and it was not until 1777 when Johann Beckmann, a professor of economics at the University of Göttingen in Germany, published his book *Guide to Technology*. [30] Beckmann was the first person to teach and write about technology in an academic manner, arguing that "the rapid spread and increasing importance of the useful arts demanded that we teach them in a systemic order … [and] interconnected."[31]

What happened is that the ancient notion divided the arts into two major categories: the fine arts and the practical arts. The purpose of the fine arts was to adorn and entertain; these include crafts, painting, sculpture, theater, and music, among others. The practical arts were those that produced goods and tools for everyday use and for the practice of a profession; disciplines such as architecture, chemistry (or its predecessor, alchemy), metallurgy, masonry, manufacturing, and law fell into this category, to name a few.

Technology encompassed both the fine arts and the practical arts. It was not until Beckmann published his book that a distinction began to be made between the two, to the point where in our times, when we talk about technology, we almost exclusively associate it with the practical arts.

Technology encompassed both the fine arts and the practical arts.

Cultural anthropology, a branch of the study of human beings that focuses on culture, defines technology as the human way of interacting with the material world or "the most obvious feature of a culture." [32] This definition is appealing because it places technology in the perspective of a broad and interconnected system, such as culture. Technology is not only material but also cultural. It exists at the intersection of what is known as the material culture of peoples.

Cultural anthropology studies human groups following a pattern of analysis that includes five elements present in every culture:

1. *Kinship*: the observation and study of interpersonal relationships within the family framework.
2. *Society*: the observation and study of interpersonal relationships outside the family framework, with the rest of the members of the immediate social group.
3. *Economy*: the observation and study of the relationships between human beings and their natural environment for the use and utilization of resources.
4. *Politics*: the observation and study of the relationships that govern the social group and its interactions with other social groups.
5. *Religion*: the observation and study of human beings in relation to their worldview and their relationship with God, the afterlife, and invisible phenomena.

Technology affects and is affected by each of these five areas. This brings us much closer to a more holistic and integrated concept of technology.

Technology belongs to culture. In fact, as we saw in the previous chapter, the biggest obstacle in adopting any technology is the cultural resistance it may generate. Installing something is easy; changing culture is difficult. Technology requires not only purchasing the right devices but also creating the right culture.

Technology encompasses countless intangible things. In fact, many of the technological transformations that humanity has undergone throughout history belong to the realm of ideas. Language and its writing, for example, are not objects. Software is not either. At this moment, not even the money in your bank account. We can extrapolate this truth to include Christian doctrine. Am I saying that doctrine is technology? In this broad, non-reductive sense, yes. It belongs to religious culture. Doctrine is an ideological tool that shapes our worldview and allows us to interact with the material world. And this does not make it any less important. On the contrary, it is a tool that has transformed the course of history.

Technology encompasses countless intangible things.

The church as an organization is technology. The church is a "holistic and integrated system" or a social organism - borrowing the words of Kevin Kelly that open this chapter - with relationships, ministries, dogmas, economy, and religious practices that provide viable alternatives for seeing and organizing our environment.

The church cannot "get into that technology stuff" because it has never left it. And this reality of the church as technology places it at the heart of this book. If the church is called to have a ministerial impact, and the church as a broad concept is technology, then technology is called to have a ministerial impact.

What Is Technology?

Knowing then that technology is a holistic system that affects all areas of culture, I propose this practical definition of technology informed by cultural anthropology:

> Technology is an interconnected cultural network of people, systems, processes, policies, and devices that enables us to interact with the world around us.

I ask you to take a few minutes to absorb this definition. Read it again. In fact, memorize it. This definition has profound ministerial implications that I would like to explore further.

People

Technology is by the people and for the people. A technology strategy cannot be established in any organization without starting with the people, both those who will lead it and those who will consume it.

The Bible prioritizes people in technology because the one who uses technology is more important than the technology itself:

> Does the ax raise itself above the person who swings it, or the saw boast against the one who uses it? As if a rod were to wield the person who lifts it up, or a club brandish the one who is not wood! (Isaiah 10:15).

This is good news for the church. The church is the congregation of those called to follow Christ, which means that the church is not the building but the people. It exists by the people and for the people. Just as we do not choose just anyone to be the pastor of the church, we should not choose just anyone to carry out its technological plans.

Having a selection criterion is imperative. It is expected that the pastor or minister is a person prepared and trained in matters of faith and doctrine, of the church and Christian practices. In this sense, the pastor is a technologist of the tasks of church ministry, a trained technician in ethics, exegesis, homiletics, and apologetics. Similarly, the personnel in charge of any ministry technology initiative must be capable and have the necessary training.

A technology strategy cannot be established in any organization without starting with the people.

Having a training and development plan is equally imperative. Just as we do not start a church without having designated and committed leadership, we should not initiate a technology implementation without having responsible and committed individuals for it. It is necessary to have a principle of accountability, because without accountability, there is no blessing, to give an account and manage the resources available.

We need the right people to plan, design, implement, and maintain the technology. A side note: No project should overlook the need for personnel in the operational and maintenance phase. Many technology projects end up abandoned due to neglect after implementation, simply because long-term maintenance was not properly planned.

But where do we find these people? Of course, the answer depends on many factors, but there are at least four ways to find qualified personnel:

1. Leaders,
2. Employees,
3. Volunteers and
4. Consultants.

As a technology manager, I must balance these four sources of personnel frequently. Each group has different characteristics and contributions. Each group has its own budgetary needs as well.

I rely on my employees for responsibilities related to critical systems, medium and long-term projects, and regular maintenance. I cannot rely on volunteers for these tasks because the time they can dedicate, while appreciated, is irregular and sporadic. Do not think that volunteers are free because you do not have to pay them; that mindset is incorrect. It will cost you something. You invest a significant amount of time (and as the saying goes, time is money) training them before you can delegate tasks that require greater commitment or care. You must also consider the potential risk and cost of errors, as well as the opportunity cost of having to delay a project because we depend on volunteered time. That is why I assign short and even trivial tasks to volunteers, allowing me to free up the effective time of employees.

When I do not have the necessary experience or knowledge for a particular project, or when the combined time of my workgroup is fully occupied, it is my responsibility to look outward and hire consultants. Consultants complement the work for a predetermined finite time and transfer the necessary knowledge to the group so they can continue with maintenance tasks.

I provide tools for hiring a good technology advisor in **Chapter 10** of the book *Cybermaturity*, based on the story of Jethro and Moses in Exodus 18. [33]

When establishing a ministerial technology plan, it is important to build a team of personnel who can carry out the vision. This should be a formal process. There is nothing wrong or offensive about requiring each person working in this area to submit a resume and go through one or multiple interviews.

Once the workgroup is formed, regular meetings should be established, whether weekly or biweekly, for example, to keep everyone connected and accountable. I also recommend having individual meetings with each person. Plan and document each meeting thoroughly. And importantly, let us not forget to take time to celebrate achievements at the conclusion of projects.

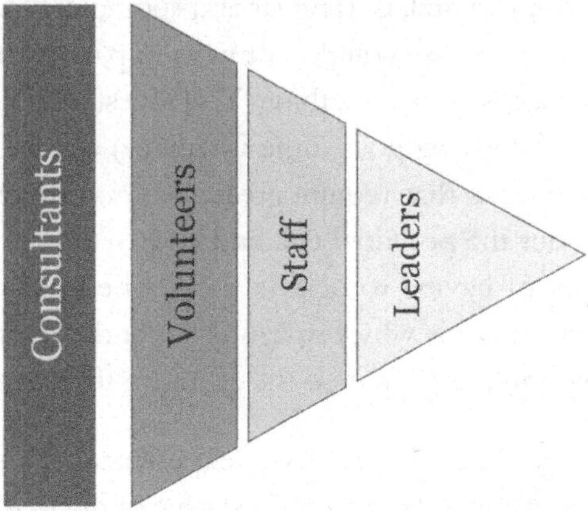

Figure 3: People

Systems

Often, we associate systems with computers, but a system is more than that. According to the Merriam-Webster dictionary, a system is "a group of devices or artificial objects or an organization forming a network especially for distributing something or serving a common purpose." [34]

Every computer is a system, but not every system is a computer. The computer is composed of various components such as memory, processor, and hard drive. It also includes a series of ports that facilitate communication with other devices called peripherals, which include the mouse (that does not want to be a lion), keyboard, monitor, and printer. These components find a particular order in the integrated circuit that allows them to interact with each other.

The collective contribution of these components is to achieve the goal of processing software applications at high speeds. The applications, in turn, serve a purpose, whether it is writing, calculating, communicating, or even playing.

Using the computer as an example helps us understand what a system consists of: Every system operates certain objects in a specific order to achieve a particular objective. This is illustrated in **Figure 4**.

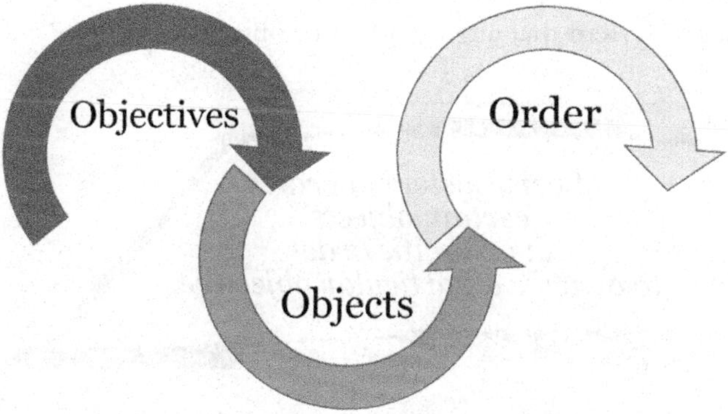

Figure 4: Systems

The design of any system should begin with the objective in mind. In other words, the system runs from back to front, but it is designed from front to back. I use the word "design" deliberately because it implies taking the time to think and plan.

When starting, we should ask ourselves, "What are we trying to achieve?" Your objective will determine the resources you will utilize, not the other way around.

Having a projector in the church does not dictate that you have to project verbatim everything you want to say. If someone visits you and asks, "Why are you using a projector?" I do not think your answer would be, "Because I have a projector." It is not about just having a projector for the sake of having one. But if your objective is to guide others in worship, a projector, along with other techniques, can help you lead the congregation. Ask, "What do I need to stimulate worship?" instead of "What can I do with this projector?"

I have noticed that in some ministries, technological decisions are made without considering the concept of a system. Do not engage in social media without knowing what you want to achieve and what kind of content you want to communicate. Do not just create a website without knowing its purpose. This practice is inadequate.

Take the time to plan and design your technology initiatives as part of a holistic system that aligns with your objectives.

Every system operates certain objects in a specific order to achieve a particular objective.

Many technology initiatives fail because they lack a clear objective and do not prioritize that objective properly. Write your objective in the most concise and detailed manner possible. Put the objective first. Regularly review your initiatives considering that objective. Make it a necessary and routine process.

The objective guides the objects. Only when you have a defined objective can you ask, "What do I need to achieve it?" I will discuss this topic in more detail later when talking about "Devices."

For now, I want to emphasize that objects should not be chosen randomly. Do not buy the first thing you see or the most attractive item or what is on sale. When acquiring devices, think of the entire system. Conduct a comparative evaluation, have a return-on-investment mindset, and consider the total cost of ownership. [35]

There are different ways to accomplish these three things, ranging from empirical methods to professional and formal analysis. Many companies that offer technology services may already have this information available. And if they do not, you can always request it before deciding.

When acquiring devices, think of the entire system.

Many ministries have asked me to help them establish their digital presence. We embark on the process of domain registration, selecting cloud hosting and programming platforms, to produce the initial content. It is common for all the costs of the startup investment to be covered, but when it comes time to renew the registration and hosting services, or to keep the website content updated, the initiative is abandoned. Every technology has immediate and long-term costs. You must consider the total cost in balance with the return on your investment.

With a clear objective and aligned objects, define how they are organized and related to each other. An overhead projector needs a screen in front, a stand, cables of a certain length, adapters to connect different audio and video sources, a signal switcher, a character generator, a computer, which in turn has its own needs.

It is likely and even recommended that front monitors are also needed. It is very uncomfortable for presenters to have to constantly turn around to see if what is being projected on the screen matches their point. The control and production area should be appropriate and have visibility to the sanctuary.

When Project "R" started using the equipment they acquired, the control room was closed off and had no visibility to the sanctuary. The control staff was always lost as to what should be displayed on the screens, and the technology became a distraction and sometimes even a source of laughter.

Corporate worship is a system. God gave Moses precise instructions to build a tabernacle, and within the tabernacle, a sanctuary. I will explore this topic more thoroughly in **Chapter 6**, but for now, let us take note of the elements of this system.

Corporate worship is a system.

First, the objective of the holy place was the worship of God. Second, the holy place was to contain three specific groups of objects: a table for placing bread, a lampstand for illumination, and an altar for burning incense. And third, these objects were arranged in a particular order.

Order implies space-time. In other words, there is a space where the functions of the objects are performed, and there is a time, rhythm, and frequency at which they are carried out.

Exodus 39:37 tells us that they brought to the tabernacle, among other things, "The pure candlestick, with the lamps thereof, even with the lamps to be *set in order*, and all the vessels thereof, and the oil for light" (KJV, emphasis added). The same goes for the table: "And thou shalt bring in the table, and *set in order* the things that are to be *set in order* upon it" (Exodus 40:4), "And he set the bread *in order* upon it before the Lord; as the Lord had commanded Moses" (Exodus 40:23).

While the utensils may have changed in the worship of the church, the objective remains the same and the principle endures. Order is an order: "But everything should be done in a fitting and orderly way" (1 Corinthians 14:40).

For an effective use of technology in ministry, we need to think systematically. This means defining the objective, the objects, and the order, and regularly reviewing the components of the system to ensure they still serve their purpose. Make sure the objective remains valid, the objects are still appropriate, and they are arranged in the best possible way.

Order is an order.

Processes

A process is a series of successive steps to be followed with a purpose. It literally means to move forward, hence the word "procession" as in religious processions. A process involves moving data from one place to another to generate information or moving objects from one state to another over time, such as in a production line, for example.

Processes bring people together with systems. They consist of the tasks that people perform in using the system with the intention of producing something. Processes produce results.

Effective technology processes are also finite, meaning they have a beginning and an end. In fact, a poorly programmed process in a computer is what causes it to "freeze" when it enters an infinite cycle or loop. Thank God for the power switch. It gives us the ability to terminate an endless process that consumes system resources. The analogy with ministerial processes is very valid. If an issue drains the church's resources, stresses the pastor, and seems to have no solution, it is likely that the best course of action is to interrupt it.

According to Dan Madison, "Process experts claim that 85 percent of all [organizational] problems can be attributed to their processes. The other 15 percent is due to people." [36] This can be enlightening for a conflict-ridden church. It is highly likely that they have problematic processes rather than problematic people. Madison further explains that when there are problems in an organization, it is common to try to identify a culprit, passing the blame like a hot potato among several individuals until the issue is forgotten. Rarely do we stop to consider that the process, not the person, could be inadequate or inefficient, and we do not focus our efforts on adjusting the process or pressing the interrupt button to resolve the situation.

Having a process mindset is crucial when using technology in ministry, whether it is existing or new. Business Process Mapping (BPM), which involves mapping, improving, and managing business processes, has given rise to a whole discipline of study, produced relevant literature, and created a specialization in business administration, along with many software applications that allow the visualization of task flow, data flow, and information flow. By the way, flowcharts are the preferred tool for documenting processes.

Continuing with our example, the content to be displayed on a projector during the worship service must go through a process. The process involves several individuals responsible for different tasks, certain data and files that need to be gathered, and specific information that needs to be shared until the final presentation is achieved.

At the same time, it is important to note that the process connects each key person with various systems to complete their tasks. These systems include the projector, but also sound equipment and tools for Bible study, among others.

Figure 5 on the next page presents a possible task map. Each of these tasks can be represented in a flowchart. You may be thinking that documenting a process is challenging, especially if you must learn how to create flowcharts. But it does not have to be complicated.

People	Tasks
Worship Leader	Chooses songs for worship service
	Rehearses with singers and musicians
	Approves the music for service
	Sends lyrics and instructions to the Control Panel
	Schedules the dressed rehearsal and sends invitations
	Leads worship during service
Worship Team	Rehearse with Worship Leader and musicians
	Participate in dressed rehearsal
	Participate in worship during service
Control Panel	Receives information from Worship Leader
	Receives information from the Service Lead
	Receives information from the Pastor
	Compiles information and publishes a consolidated presentation
	Confirms that the presentation is ready to go
	Tests the presentation during the dressed rehearsal
	Controls the presentation during service
Service Lead	Chooses opening Scripture
	Collects all the announcements for the week
	Creates the service outline
	Gets service approval from the pastor
	Participates in dressed rehearsal
	Leads the service
Pastor	Prepares sermon
	Chooses content to be included in the presentation
	Chooses Scripture to be included in the presentation
	Sends content and instructions to Control Panel
	Participates in dressed rehearsal
	Approves service
	Preaches sermon during service

Figure 5: Work Map

You can follow the following simple yet practical format to create a flowchart. If you review **Figure 5**, you will notice that some tasks can be defined as activities, others as decisions, and some require data storage. These types of tasks are graphically represented by a rectangle, a diamond, and a bullet point, respectively, and connect with arrows, as shown in **Figure 6**. Note that the decision diamond has two possible outcomes and two outgoing arrows to represent Yes or No.

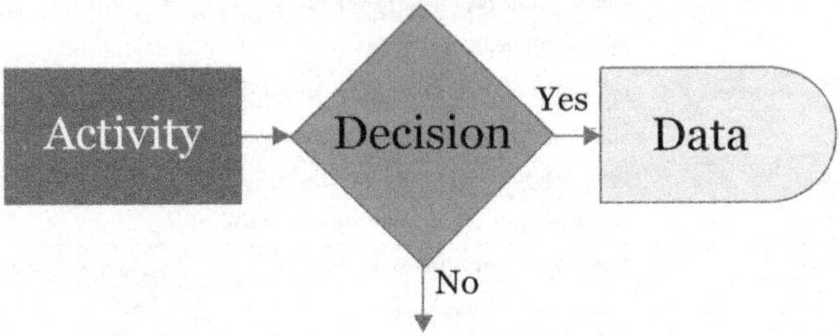

Figure 6: Workflow Diagram

In typical process documentation, the task map from **Figure 5** is combined with the task diagrams from **Figure 6** to form what is known as a swim lane chart.

This type of documentation adopts the metaphor of a swimming pool divided into lanes. The entire pool represents the process, while each lane represents the roles that individuals play. In smaller organizations, it is likely that one person will assume multiple roles.

Under this matrix, activities, decisions, and data are organized at different points along each swim lane. The arrows not only connect activities with decisions and data, but also indicate the direction in which activities flow.

The list of activities from **Figure 5** in this process would then be transformed into something like the swim lane chart shown in **Figure 7** below:

Please note that if an activity involves multiple people, it can cross multiple swim lanes, which is not allowed in swimming but is permitted in process diagramming. This is the case with the "Dress Rehearsal" and "Worship Service" activities. Also, note that in this example, the "Publish Presentation" activity produces data, in this case, a presentation file, which is why the bullet shape is used instead of the rectangle. Lastly, you can be as general or specific as desired; some of these activities could be further subdivided into multiple smaller activities. For example, "Musicians" could be moved to their own swim lane.

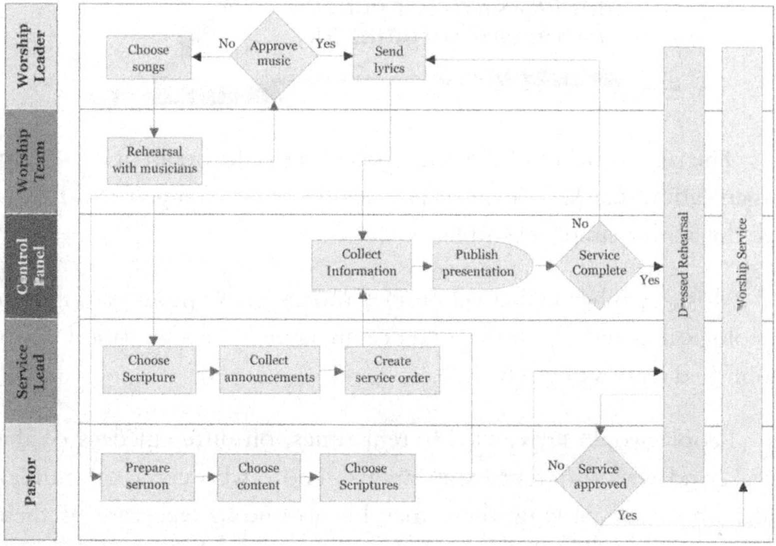

Figure 7: Workflow Swim Lanes

For effective use of technology in ministry, it is essential to define and document the involved processes. This means outlining, with careful attention to details, how people interact with systems and perform their various tasks, make decisions, and store information.

When processes are documented in this visual manner, the ministry team stays well-informed and can easily identify bottlenecks or delays in the process. Armed with this information, process improvement and even reengineering can be fostered.

Policies

Policies are guidelines, rules, or procedures that govern people's actions and their interactions with systems. In other words, they provide context to processes to ensure coordinated and efficient functioning.

An organization that lacks explicit policies becomes disorganized.

The lack of clear policies and procedures is the origin of chaos. An organization that lacks explicit policies becomes disorganized. Let us see this with a simple example.

Most churches gather on Sunday mornings. Without getting into theological details about this practice, imagine if a congregation never published its meeting schedule.

People would arrive at different times, on different days of the week, and with different expectations. Those who come on Sunday might not find parking, there may be no one to take care of their children, and when they enter the sanctuary, the lights are off. The pastor went golfing with friends that day, and the musicians are each playing a different song. Total chaos.

On the other hand, if the Sunday gathering is stated to start at 10:30 am, it is much more likely that other things will align to produce a better experience. So, "Sundays at 10:30 am" is a policy, and this policy governs the practice and procedures of that congregation - which I mentioned was imaginary, any resemblance to reality is purely coincidental.

Policies guide processes. Going back to the swim-lane diagram in **Figure 7**, we can now connect this policy with the process. If we combine the activity "Worship Service" with the policy "Sundays at 10:30 am," we can trace the rest of the policies by following the diagram in reverse.

We could say that the "Dress Rehearsal" must be before that time, for example, "Saturday at 8:30 am" - hopefully with breakfast included; that the Control Panel must "Publish Presentation" before Friday night in the central storage, that both the Worship Leader, the Master of Ceremony, and the Pastor must provide the necessary information to the Control Panel before Thursday morning, and that the "Music Rehearsal" is on Tuesday night.

These time policies could even be written at the top of **Figure 7** as column headings for the activities.

The worship service will not be any less spiritual if we plan it well and adhere to certain governing policies. The desert tabernacle had its own policies. Animals for sacrifice had to be without defect, the oil had to come from crushed olives, the lampstand and incense could not cease burning, and the anointing oil was not a perfume for common use, among other things.

Dante Gebel puts it this way: "Planning is not only not a sin but, on the contrary, it is a determining condition for God to do something serious with our lives." [37] And he continues fervently: "[we] have made an art of improvisation and mediocrity, except that since those words sound too strong, we call them 'the leading of the Spirit'." [38]

Defining precise policies and procedures does not limit the action of the Holy Spirit, but rather provides a formal context for His manifestation and blessings to reach as many people as possible.

In general, when it comes to technology policies, at least four types of policies and procedures need to be outlined, with the overall objective of providing a framework and even legal protection:

- *Acceptable Use Policies*: These are the rules that govern the acceptable use of technology. Many information systems require explicit or implicit approval of usage policies. Both the policies and usage procedures should be created with as much detail as possible.

- *Control Policies*: These policies pertain to the governance of the organization, including the role of the leadership body, user requirements, resource usage, finances, and how technology systems support these activities.

- *Security Policies*: There are various security protocols that encompass physical, industrial, and information system security. The SANS Institute for Systems, Networks, and Security Administration, [39] for example, provides valuable templates that can be tailored to the context and needs of any organization.

- *Data Management Policies*: These policies establish guidelines for the collection, storage, retention, and disposal of data within the organization. They address issues such as data backup procedures, data retention periods, data classification, and compliance with data protection regulations. These policies ensure that data is managed effectively, securely, and in compliance with legal requirements.

Every Christian organization must establish its own policies for the use, control, security, and data management of its technological resources to ensure their proper adoption and employment.

When drafting the policies for use, control, and security, make sure to include not only the rules of the game but also the players and their roles. The policies regarding the people involved should be detailed enough to include qualifications, responsibilities, functions, and expectations.

For example, when defining the policies for the use, control, and security of worship systems, it is necessary to clearly define:

1. *Qualifications*: The calling, skills, and talents of individuals, the required knowledge to maintain and manage the systems, and the experience they bring to the rest of the team.

2. *Responsibilities*: Those responsibilities for which qualified individuals are held accountable, such as leading worship during the service or training members to participate in congregational services.

3. *Functions*: Think in terms of day-to-day operations, rehearsing the music, singing the songs, adjusting the sound, projecting the lyrics, and preaching the sermon are all functions of the individuals assigned with those responsibilities.

I want to share one more tool. It is called the RACI matrix. [40] The RACI matrix is a visual way to align responsibilities with roles and accountable individuals. RACI is an acronym that stands for:

R responsible people for completing a specific task or activity.
A accountable people, ultimately responsible for the outcome.
C consulted people providing input for decisions and actions.
I informed people about progress, decisions, or outcome.

Continuing with the example of the worship team, a matrix can be created with the following roles: musician, vocalist, technician (sound engineer), producer (audiovisual media producer - familiar with using the projector), and pastor.

These roles intersect with the functions to create the matrix, assigning one or more of the four RACI letters to determine responsibilities, as shown in the following figure:

Function	Musician	Singer	Technician	Service Lead	Pastor
Rehearse the music	RA	A	A		I
Sing the songs	A	RA	A		C
Adjust the sound	A	A	RA	I	
Project the lyrics			A	R	
Preach the sermon	I	I	I	I	RA

Figure 8: RACI Matrix

Note that not all roles are assigned in the matrix, such as the pastor, not being involved in adjusting the sound, and any role may have multiple responsibilities, such as being responsible and accountable for playing the music.

Policies and procedure manuals should be part of the organizational documentation, in other words, contrary to the practice of many Christian organizations that lack a documentation culture, it is important to take the time to sit down and write them and share them in a common and accessible place for everyone.

Devices

What people commonly refer to as technology are the devices or gadgets. This can be understood since devices are the tangible objects of technology that we can see and touch.

At this point in your reading, I am convinced that you already have a much broader concept. But if not, let me illustrate it with a current example: your smartphone.

The device consists of layers of physical components: the aluminum chassis and the silica glass, perhaps protected by an impact-resistant cover, form the outer layer. Then come the interfaces that allow interaction with the device, typically a series of buttons and ports that facilitate basic input and output (I/O) functions. Inside this shell, there is the power source or battery and the integrated circuits that move information through electrical currents.

But the phone, no matter how smart it is, is not useful without the thousands of invisible developers who program its operating system and millions of other applications. Then there are the invisible radio and microwave waves that make cellular communication and data transfer possible. Next, there is the interconnected app store that allows the download of music, videos, and books. Additionally, you can obtain many complementary or peripheral devices, such as the camera (I), headphones (O), and charger (I).

Following this logic, your smart device is a system. It encompasses a range of combined technologies, along with people, systems, processes, and policies, making it a complex tool of convergence.

In the implementation of any technology, multiple people can participate, multiple systems can be coordinated, multiple processes can be defined, multiple policies can be established, and multiple devices and materials can be used, which in turn can utilize various auxiliary and complementary objects.

You must have criteria and a selection policy when acquiring devices for your ministry. The evaluation should consider both the device and its manufacturer. Since technology is a cultural system, it is essential to evaluate the device manufacturer to safeguard the investment and ensure that it delivers the promised benefits within the expected timeframe. For example, a projector should last five years and not six months, and if it becomes damaged at six months, the manufacturer should not have gone out of business and should be able to honor the warranty or provide appropriate repair services.

Figure 8 organizes some of these criteria to consider, in two columns corresponding to the device and its manufacturer. There may be other considerations, but your ministry team can use this example to generate its own criteria.

Devices	Manufacturer
Features and durability	Longevity
Ease of use	Future strategy
Total Cost of Ownership	Distribution channels
Architecture	Business reach
Required accessories	Geographic location
Industry standards	Industry position
Maintenance requirements	Technical support

Figure 9: Device Selection Criteria

One possible way to evaluate is by assigning a score to each of these criteria on a value scale, let us say from 0 to 5, and adding up the scores for various candidates in a matrix. The goal is not to achieve an exact measurement, but rather one that can assist us in decision-making.

The model can be adapted to evaluate less obvious devices: furniture, hymnals, Bibles, software, or children's curriculum. The same approach can be applied to service providers.

Thus, our definition of technology is illustrated with **Figure 10** below:

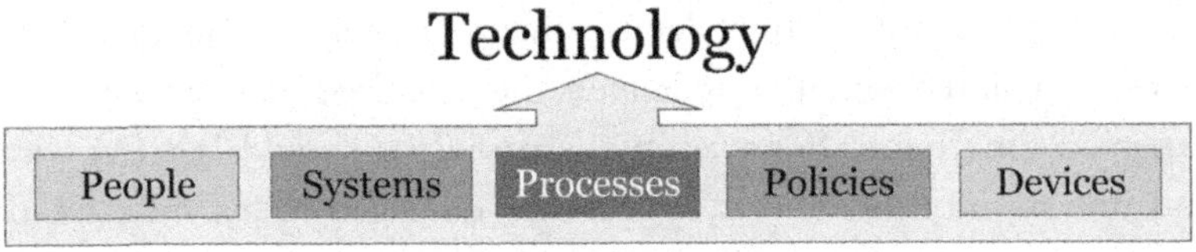

Figure 10: Technology Definition

This concludes our practical definition of technology. I hope to have expanded your understanding of technology by revealing its five components and the practical implications that stem from them. Now let us examine the three fundamental activities of technology.

TECHNOLOGY TRILOGY

The prophecy will be fulfilled soon, but before it can be, the Oracle must be consulted.

– MORFEO, THE MATRIX RELOADED.

In war, then, let your great object be victory, not lengthy campaigns. [41]

– SUN TZU, CHINESE GENERAL AND STRATEGIST.

I was working at Fuller Theological Seminary when they premiered the second film in the Matrix series on the big screen. The central theme of this science fiction series is the liberation of humanity held captive by machines. The machines survive by extracting energy from humans, who are connected to sensors and cables in an extraction farm. The machines have created software that simulates life on Earth and replaces the human experience of slavery with an alternate virtual reality. This computerized system is called the Matrix.

Like any computerized system, the Matrix has a margin of error of 1% of humans for whom the software does not work. This remaining 1% is tasked with freeing the rest. They have learned to enter and exit the Matrix at will, but only one person, Neo, the One – played by Keanu Reeves – can manipulate the code to dismantle the Matrix from within. If successful, he would free every human from the Matrix and bring them into the new city of Zion, which was protected by a fleet of ships.

These movies are filled with spiritual themes such as salvation, redemption, the struggle between good and evil, and the fulfillment of prophecies of a new messiah who can challenge the Matrix and ultimately destroy it, leading every human to awaken to the real world and freedom.

In The Matrix Reloaded, the second film in the series, there is a scene where Morpheus – played by Laurence Fishburne – a man who has already been freed, arrives late to a meeting of the generals of the Zion ships. The machines are preparing a massive attack against Zion to destroy all the rebellious humans, so the commander has ordered everyone to return to Zion to defend the city.

Morpheus suggests that they stay behind to re-enter the Matrix and consult the Oracle. The Oracle is a software program that resides in the Matrix but sympathizes with humans and helps them understand, by scanning their programming language, the fulfillment of prophecies. She is an enigmatic kind of prophet who points out the weaknesses of the Matrix in the virtual world so that they can overcome it in the real world.

When the second movie premiered, I was working as a database administrator, maintaining the student and finance management system of the seminary. This system, used by nearly three thousand universities worldwide, runs on a software platform provided by the company Oracle.

When I returned to the office the following Monday after the premiere, I replaced the standard nameplate with my name and title with a photo of myself wearing sunglasses, manipulated to make me look like Morpheus, and the phrase: "The Oracle must be consulted." I used the Oracle logo instead of the word.

This caused a sensation. For the following weeks, employees and students of the seminary would come to my cubicle to see my sign. It was a lot of fun. They made all sorts of comments. Eventually, I became known as "the Oracle." Whenever they needed any information from the system, they would say, "I am here to consult the Oracle."

A Good Trilogy

The film industry has produced several successful trilogies. These series include Jason Bourne, The Dark Knight, Toy Story, The Lord of the Rings, The Chronicles of Narnia, Star Wars (Episodes IV-VI), and The Godfather. This storytelling format in three acts comes from classical Greece and is known by its Greek term "trilogy," meaning a work in three parts. Matrix is one of my favorite trilogies.

A trilogy tells a complete story. It follows the structure of a well-written piece: introduction, development, and conclusion. The introduction presents the general theme, the main characters, and the problem to be solved. It sets the purpose and direction of the story, captures attention, and fosters interest. The development builds up the complexity that needs to be unraveled and increases the expectation of the outcome, in other words, the development leads to the conclusion. At the conclusion, the problem is resolved, and the whole story makes sense.

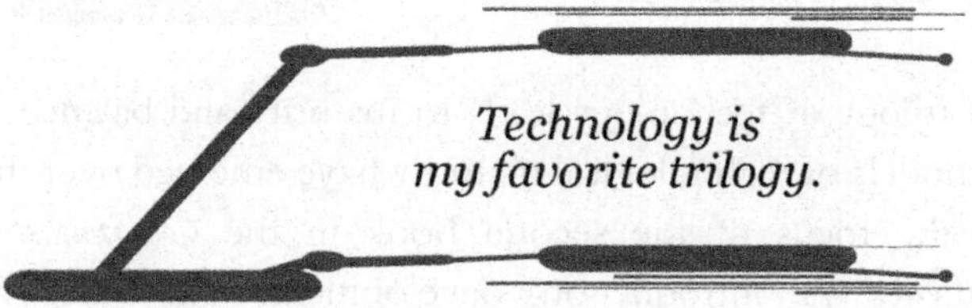

In other words, a good trilogy matures with each episode. There is a progression towards the maturity of the story. As it approaches the climax, the story gains more substance and meaning. It has more to offer. It becomes more robust. Characters mature, plans yield results, situations are resolved, and the war is won. It moves towards balance and a moral lesson to bring a sense of satisfaction and fulfillment.

Technology is my favorite trilogy. Technology tells its story in three acts and follows the structure of a good trilogy to present a complete narrative where each portion complements the other two.

In the previous chapter, we defined technology in terms of culture and the practical application of its five components. This definition corrects the minimalist or reductionist thinking one may have about technology, which is fine as an entry point.

However, the most frustrating situations in my professional career have not arisen from a lack of understanding of "what is technology?", but rather from not knowing "what technology is for?" The trilogy of technology answers this second question. If we do not understand the tripartite nature of technology, we will never be able to fully leverage its potential. Our benefit will be incomplete because we will be unaware of its functions.

The trilogy of technology will promote the growth of your ministry.

The trilogy of technology leads to maturity and balance. In fact, several models of technological maturity have emerged over the years. That is the focus of the second book in the *Cyberministry Series*. Furthermore, the introduction, development, and conclusions of technology, all three, should facilitate the progress of the mission of every Christian organization.

Only when we understand the trilogy of technology can we make our technological initiatives successful rather than failures, and with their success, the success of the mission is encouraged.

The trilogy of technology will promote the growth of your ministry. It will first help you understand so that you can grow in wisdom and avoid the failure of your programs (Proverbs 15:24).

Technology Trilogy

The trilogy of technology consists of strategy, tactics, and intelligence. I refer to these as the three fundamental activities of technology. We can think of strategy, tactics, and intelligence as the introduction, development, and conclusions of technology.

The purpose of technology is fully realized and matures to its maximum potential when strategy, tactics, and intelligence are carefully considered. From these fundamental activities, all subsequent activities of any technological initiative are derived.

> *The trilogy of technology consists of strategy, tactics, and intelligence.*

Over the years, I have encountered various ministries and Christian organizations that focus primarily on tactics. They constantly organize or participate in numerous events, utilize various available media and even the latest technology, but they fail to demonstrate the results of all their activity. Ultimately, if they even manage to assess their efforts, they do not experience growth in numbers or engagement, for instance. Their members are not more committed, and they fail to make a significant impact in the community. It seems much easier to have many activities and give the impression of busyness than to take the time to develop an action plan.

As we will see, we cannot underestimate the time it takes to develop such an action plan. The resources invested in strategic planning will never be wasted, even if they do not immediately address pressing needs.

Strategy, tactics, and intelligence are the three allies of technology. All three must work in perfect coordination to achieve the appropriate results in ministry. Once the strategy is defined, tactics translate it into practical implementation, and intelligence analyzes the effectiveness of the tactics in achieving the strategy and provides feedback, thus closing the cycle, as shown in **Figure 11**.

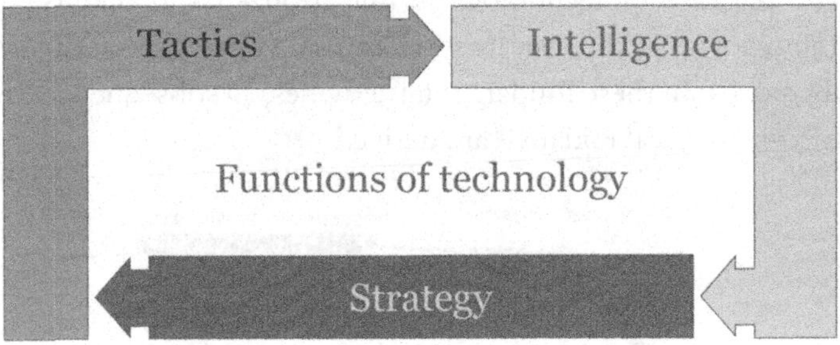

Figure 11: Technology Fundamental Functions

Let us now review some specific considerations for each of these fundamental functions or activities.

Strategy

This model of strategy, tactics, and intelligence originates from the military context. The Bible itself places strategic thinking in the military framework. The book of Proverbs encourages us to establish a good strategy before engaging in tactical matters of warfare: "Plans are established by seeking advice; so if you wage war, obtain guidance."

(Proverbs 20:18), and also: "Surely you need guidance to wage war, and victory is won through many advisers" (Proverbs 24:6).

Strategy defines the desired victory, the direction to be taken, and the tactical options. As Proverbs states, strategy is created through counsel, involving key individuals who design and execute it. The available technology informs and influences decisions on the battlefield. It is no coincidence that much technological development originates from the needs of the battlefield.

The word "strategy" comes from the Greek word *strategós*, which is a combination of *stratós* - army or host, derived from the verb *strateúo* - to serve in the military or engage in warfare, and *ágo* - leader or commander. [42]

Luke refers to the chief captain of the temple guard (Luke 22:4, 52, and Acts 4:1; 5:24, 26) and the Roman magistrates (Acts 16:20, 22, 35, 38) as *strategoi*, who were responsible for both Jewish and Roman troops or *stráteuma*.

Paul also employs these words liberally as a metaphor for the Christian life. For Paul, the believer is a soldier of Jesus Christ (2 Timothy 2:3-4; 1 Corinthians 9:7), who combats with the armor of God (Ephesians 6:13), fights according to the Spirit (2 Corinthians 10:3), and employs the powerful weapons of our warfare – *strateías* (2 Corinthians 10:4).

This metaphor was further expanded by the Apostolic Fathers, particularly Clement, who wrote: "Let us then, men and brethren, with all energy act the part of soldiers, [*strateusómetha*], in accordance with His holy commandments" (1 Clement 37:1). [43]

Strategy is the responsibility of leadership. Strategy guides us in the spiritual battle as soldiers of God's army. Therefore, strategy is essential in determining the success of any ministry initiative. As leaders, it is our responsibility to think and act strategically, both in general terms and in terms of the strategic use of technology in ministry.

Strategy guides us in the spiritual battle as soldiers of God's army.

Strategic Planning

The field of business administration, as a discipline of study, has outlined how to create a strategy through the concept of "strategic planning."

Strategic planning begins with a process of organizational and cultural introspection that culminates in a document with the following hierarchical order:

- *Vision:* The vision is visual, to put it redundantly. It is a detailed picture of the future we want to see. It does not express our current state but our aspirations. It does not need to be too short; it is okay if you write the vision in a couple of pages with descriptive details.

- *Mission:* The mission is the performance of a task. Think of it as a map of the present to move towards the future. It is a manual that describes what we should keep doing. These are the key activities. If the vision is the destination, the mission is the route. To define your mission, provide a brief description of what you do daily.

- *Values:* Values are the irrefutable principles in the exercise of the mission. Values are precedent decisions - decisions made beforehand. Think of them as the rails of a railroad; you can take different routes and change the speed, but you never derail. They are the faith that drives us; the motivation behind what we do. To define your values, specify three to five values, never more than seven, and if possible, use an acronym to make them easy to remember.

- *Strategies:* Strategies are the overall objectives that, by fulfilling the mission, will allow you to achieve the vision. Think in terms of the mode of transportation you will use. You can go by donkey, train, or airplane, but you must decide in advance which one you will use. Choose three to five strategies to work on. Remember that less is more, meaning it is easier to achieve what you propose if the list is short.

- *Goals:* Goals are action. They are the journey. Think of them as milestones along the way. You may have heard of the concept of "SMARTER goals," which stands for specific, measurable, attainable, relevant, tangible, exciting, and rewarding goals for each strategy that is part of your plan.

- *Methods:* Methods are the routine operations that will allow us to achieve the goals. Methods are practiced. Think of them as fueling the locomotive, changing the switch levers at intersections, or picking up the package left along the way. They are systematic procedures or techniques that are executed in practice. Although methods can be adjusted along the way, the vision, mission, values, and strategies tend to be more permanent. Write at least one method to achieve each of your goals.

- *Metrics:* Metrics are the accounting markers that show how much progress has been made over time and whether the methods are helping to achieve the goals. They are like the GPS on the map, the speed, the mileage, the temperature of the boilers, and the amount of cargo transported. As we will see later, they are your gateway to self-assessment. Define two or three performance indicators for each strategy.

The vision, mission, and values belong to the realm of strategy, strategies, goals, and methods belong to tactics, and metrics belong to intelligence. Note that strategic planning considers the three fundamental activities of technology.

There have been some attempts to apply strategic planning to Christian ministry, especially the extensive work of Aubrey Malphurs, particularly his book: *Advanced Strategic Planning: A 21st Century Model for Church and Ministry Leaders.* [44]

There have also been some critics of strategic planning. Will Mancini, for example, warns about the limitations of traditional strategic planning in the context of ministry. In his book, *Church Unique: How Missional Leaders Cast Vision, Capture Culture, and Create Movement*, Mancini highlights three fallacies of strategic planning: the complexity fallacy, which explains that planning can become so complex that it loses clarity; the responsibility fallacy, where each working group is so focused on showcasing their individual results that they lose sight of the bigger vision and team synergy; and the predictability fallacy, which assumes that the medium-term future will be the same as the immediate past, when in reality the world is rapidly changing, requiring flexibility and quick adaptation. [45]

Those who oppose strategic planning argue that the process tends to be lengthy and cumbersome, and that the resulting documentation is overly extensive, with a list of goals, strategies, and methods that become difficult to follow. They also claim that there is an excessive emphasis on defining quantitative and qualitative measures of success, which often reduce the plan to mere numbers.

From my experience, I know that this emphasis can be disproportionate, and it is very challenging to define metrics of success in ministry. After all, efficiency does not guarantee effectiveness, and busyness does not necessarily equate to impact. We could showcase impressive numbers that demonstrate how busy we are doing something that is not essential. Someone once said that good is the enemy of the best. Efficiency is akin to Martha, while effectiveness resembles Mary.

However, it does not have to be this way. It all depends on the strategic planning process, the timeframe chosen, and the execution of the plan by skilled strategists and advisors.

A good ministry strategy is healthy when it aims for effectiveness. Aubrey Malphurs writes about this issue:

> It is essential that churches do strategic planning. We hear some voices out there that argue that churches no longer need to plan strategically. What they should be saying is that churches should no longer attempt to do and rely on long-term planning. Our world is changing too quickly for long-term planning to be effective. Strategic planning that involves thinking and acting strategically will always be useful. [46]

The speed of change in our surroundings has been accelerated by many new technologies. This backlash against long-term planning is not only coming from the church but also from other organizations.

In the business world, a similar response against traditional planning methods is being experienced. That is why new methodologies have emerged, such as Agile, DevOps, Scrum, Continuous Integration, Lean Manufacturing, and Business Model Generation (BMG). [47]

These methodologies define a much more flexible way of strategic planning, with more realistic timelines, adapting promptly to market needs, the environment, and user requirements. These models offer an alternative approach to strategizing in a rapidly changing world.

To date, there is no work exploring how to adapt these methodologies to the strategy of our Christian organizations, but they could be viable approaches when we need to be flexible and adaptable.

> *Efficiency does not guarantee effectiveness.*

In any case, if we follow our definition of technology from the previous chapter, any strategic planning methodology as a thinking model is technology. Therefore, it is expected that technology permeates strategic planning in each of its stages. The important thing for our discussion is to keep in mind that, regardless of the preferred strategic method, technology serves strategic purposes.

As we will see, technology provides means to establish and communicate the vision, develop the mission in the context of the culture, achieve the proposed goals, and use performance indicators to make decisions. If we think of the strategic plan in reverse, the measurements can be a fair representation of the efficiency of the methods, the methods can help achieve the goals and strategies, achieving the goals could bring us closer to fulfilling the mission, and if we fulfill the mission, we could reach the future we want to reach.

The seven ingredients of your ministerial strategic plan are represented in **Figure 12** on the next page.

Now, how do we develop a strategy? What are its fundamental elements? Where should we start? How does technology fit into or align with the strategy? These are questions I aim to answer below.

All Ministry Strategy Begins with Prayer

It may sound obvious, but it is a practical reminder: every ministerial strategy is born in prayer.

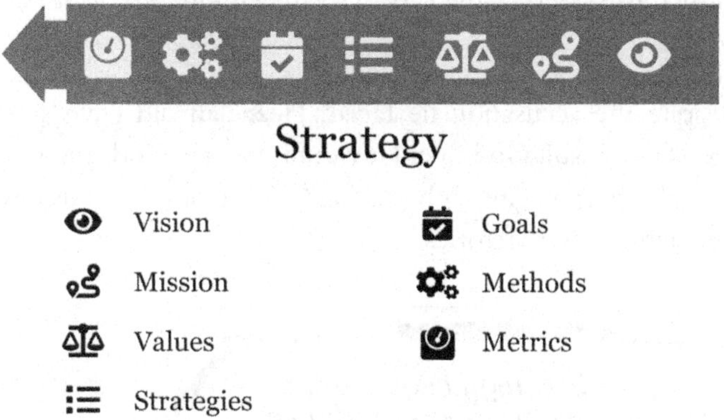

Figure 12: Ministry Strategic Plan

In 2 Kings 18 and 19, the story is told of how God delivered Hezekiah, the king of Judah, from the threats of Sennacherib, the king of Assyria. Hezekiah was one of the righteous kings of Judah. It is said that he trusted in the Lord (18:5) and was successful in all his endeavors (18:7). Sennacherib had conquered many fortified cities of Judah (18:13), but Jerusalem remained unconquered. Hezekiah decided to pay tribute to Sennacherib to avoid the downfall of the kingdom, but it was not enough.

Sennacherib then sent his field commander and chief of staff to speak to Hezekiah and threaten him, saying:

> On what are you basing this confidence of yours? You say you have the counsel and the might for war—but you speak only empty words. On whom are you depending, that you rebel against me? (18:19b-20).

So, Hezekiah consulted the Lord (19:1) and the Lord gave him a word of victory through the prophet Isaiah (19:6), who also prophesied the retreat of Sennacherib and his death by the sword when he returned to Nineveh, the capital of the empire, just as it happened (19:36-37).

Despite the accusation he faced, Hezekiah did have a strategy, which was to consult God first. He put his trust in God, and it resulted in victory. Strategy begins with seeking God's guidance to discover His plan and achieve our victories.

Strategy begins with seeking God's guidance to discover His plan and achieve our victories.

Having a strategy is not less spiritual; in fact, it is quite the opposite. A strategy rooted in seeking God's guidance and aligning our actions with His plan is a deeply spiritual approach. It demonstrates our reliance on God and our desire to follow His leading in every aspect of our ministry. A strategic mindset allows us to steward our resources, talents, and opportunities effectively for the advancement of God's kingdom. It is a way of acknowledging that our plans and actions are ultimately dependent on God's wisdom and guidance. So, having a strategy is a spiritual practice that enables us to fulfill our calling and bring glory to God.

All Ministry Strategy Must Be Written

The words of the Assyrian commander should also serve as a warning to us. We cannot claim to have a strategy if it is merely empty words without substance.

Every Christian ministry should document its strategy to give it meaning and clarity, and to allow for evaluation of its achievements. The fact that we have a written record of Hezekiah's strategy, and the outcome of his story demonstrates the need for us in ministry to write our own strategy and, when the time comes, evaluate our own results.

In the previous chapter, I established that language and writing are technology. The means of communication and recording writing are also technology. Writing the strategy inherently requires the use of technology. This is confirmed in the passage of the following point.

Every Christian ministry should document its strategy to give it meaning and clarity.

All Ministry Strategy Stems from a Clear Vision

The prophet Habakkuk, who lived a few years after Isaiah and Ezekiel, also consulted God, this time during the time of the Babylonian invasion. Habakkuk consulted God upon seeing the injustices committed by the Chaldeans.

God responded to him, saying, "Write the vision, and make it plain upon tables, that he may run that readeth it" (Habakkuk 2:2 KJV). Every ministry strategy stems from writing a clear vision.

Many times, we do not know how to achieve victory because we do not even know what victory looks like. Habakkuk wrote the vision on stone tablets; now we must write it on digital tablets. The only way to ensure that those serving in the ministry do so effectively is by defining the vision.

The vision is an image of the future we want to see. Writing our vision creates a common language of clarity, identity, and commitment that allows everyone to walk in the same direction.

Our vision of the future must consider the role that technology will play. A war strategy differs from another depending on the technology available or chosen to use. Many technological advancements have determined victory in many wars.

A newly established ministry can take this into consideration more easily. It may be more challenging for a long-standing ministry that has been walking in its vision for years because it tends to cling to outdated technologies. Remember that it is difficult to change the culture and habits of people in traditional organizations.

In any case, a strategy, and therefore a vision, is not simply a document we create to print it and store it in a file cabinet drawer. The strategy, starting with the vision, is meant to be read continuously and run with. Plan revision sessions that allow for re-visioning. Those occasions are perfect for considering technology in ministry practice.

The vision is an image of the future we want to see.

All Ministry Strategy Runs its Mission

If the vision is an image of the future where we want to go, the mission describes how we are going to run with it or how we are going to get there.

The mission is a definition of why we exist and, consequently, what we are dedicated to. I do not want to delve too deeply into the topic of the Christian mission at this moment, as I will dedicate **Chapter 5** to it, but I want to advance that the Christian mission is not just a statement but a ministerial command; the mission is not a mantra to repeat but a mandate to fulfill. In other words, the mission is not just about agreement, but about action.

This is where the Christian mission differs from the mission of a regular company: Jesus Christ told us what we must do, what he expects from us. What is recorded in Scripture defines the essence of who we are as an organization.

The mission and technology coincide in their pragmatic nature. It is in the execution of the mission where we can best align the use of technology. You devise the commission before you commission the device. That is why the ministry dictates the mission, not the other way around, and the mission dictates the technology, not the other way around. In other words, the Christian mission determines what technology should be used and how. This is what I mean when I talk about the strategic alignment of technology, which leads me to the next point.

Plan revision sessions that allow for re-visioning.

All Ministry Strategy Facilitates Communication

Your ministerial strategy is a model of communication, and this is no small thing. Nothing that exists, exists apart from language. All things, and I mean 'all' things, first existed in the intangible realm of ideas. But it was the power of the word that transformed those ideas into tangible reality.

This is a theological truth, as explained by the Apostle John:

> In the beginning was the Word, and the Word was with God, and the Word was God. He was with God in the beginning. Through him all things were made; without him nothing was made that has been made. (John 1:1-3)

This continues to be true in the world of technology. In this sense, we are also made in the image of God: in our capacity for language. Every product of human ingenuity and every technological advancement has been initiated and materialized through language. Everything continues to be created by the word, and without the word, nothing will be made.

You devise the commission
before
you commission the device.

If we want to stay committed to the mission and see the vision come to fruition, we must not only write it but also communicate it in a concise, clear, accessible, and recurring manner. When there is written clarity and intentional communication of the strategy, we will run towards its fulfillment. That is why Habakkuk continues to say:

> For the vision is yet for an appointed time, but at the end it shall speak, and not lie: though it tarry, wait for it; because it will surely come, it will not tarry (Habakkuk 2:3 KJV).

All Ministry Strategy Requires Accountability

Joshua was about to attack the city of Jericho when a warrior approached him with his sword drawn (Joshua 5:13). Joshua asked whether he was a friend or an enemy, to which the man replied, "And he said, Nay; but as captain of the host of the Lord am I now come" (Joshua 5:14). Joshua bowed down to worship him, and the prince of the army of the Lord did not stop him. Instead, he told Joshua to take off his sandals because he was standing on holy ground, just as he had instructed Moses during their encounter with the burning bush.

Commentators agree that this is a Christophany or a manifestation of the pre-incarnate Christ.

The prince of the army of the Lord – Christ – gave Joshua the vision: "See [vision], I have given into thine hand Jericho, and the king thereof, and the mighty men of valor" (Joshua 6:2). He then provided Joshua with the tactical details (Joshua 6:3-5).

What follows is remarkable. Joshua commanded the priests and the people to follow the attack plan step by step for the next seven days, without omitting anything (Joshua 6:6-19). Joshua adhered to the instructions of the prince of the army of the Lord, the priests obeyed Joshua, and the rest of the people obeyed the priests. The result was a resounding victory; the wall collapsed, and they took the city along with its spoils (Joshua 6:20). What would have happened without obedience? Without accountability, there is no victory.

Mission is not
a mantra to repeat
but a mandate to fulfill.

Let us remember that one of God's titles in the Old Testament is *Yahweh Sabaoth*, or the Lord of Hosts. It appears 283 times in the Hebrew scriptures, the first being 1 Samuel 1:3, and twice in the New Testament in Romans 9:29 and James 5:4.

The phrase implies that the spiritual angelic hosts are subject to Him, and therefore, those who worship Him in spirit and truth should also subject themselves. Christ is the supreme captain, the commander-in-chief, to whom we all, as soldiers, must submit.

The epistle of Clement to the Corinthians, mentioned earlier, continues to explore this metaphor, saying:

> Let us consider those who serve under our generals, with what order, obedience, and submissiveness they perform the things which are commanded them. All are not prefects, nor commanders of a thousand, nor of a hundred, nor of fifty, nor the like, but each one in his own rank performs the things commanded by the king and the generals (1 Clement 37:2-3). [48]

Colonel Harry G. Summers, a member of the United States Army infantry and later a war analyst, who was part of the team negotiating the Vietnam War cease fire, wrote:

> Tactical success is not necessarily strategic success, and tactical failure is not necessarily strategic failure… Ironically, our tactical successes did not prevent our strategic failure. [49]

We can excel at keeping ourselves busy (tactics) without staying focused on what is truly important (strategy). That is why strategy is paramount.

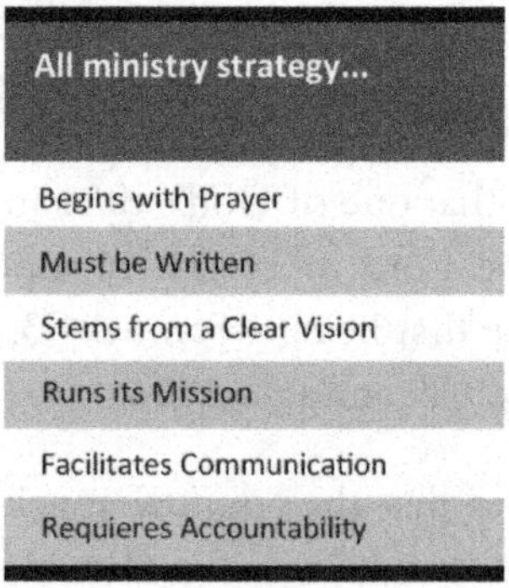

Figure 13: Ministry Strategy

Every technology initiative in ministry should be integrated at the level of the ministry's strategy, which should precede, guide, and inform our tactical actions. Let us move on to consider the second episode of our trilogy.

Tactics

Tactics are related to the everyday affairs of the organization, the management, and the operation of technology in the church ministry. While strategy answers the what (vision-mission), the why (values), and the what for (strategies), tactics answer the how, when, and where.

Tactics are more situational and depend on specific factors such as the environment, culture, available resources, or changes in leadership composition. Strategy is grounded and stable, while tactics are subject to variations of the situation and circumstances. Strategy defines the tactical course, and tactics define the practical approach.

The word "tactics" also comes from Greek, *taktikós*, and encompasses everything related to the organization of things. It is derived from the verb *tásso*, which means to arrange or order, both in the sense of systematizing or arranging and in the sense of giving commands or instructions.

According to the Merriam-Webster dictionary, tactics is "the art or skill of employing available means to accomplish an end." [50] Therefore, tactics involve the organization and necessary steps to see the strategy realized. They are the smaller-scale actions that move you towards your desired goal. The emphasis is not on planning but on the execution of the plan.

When the Persian emperor Darius decided to organize his government, he appointed three governors or ministers and placed them in charge of 120 satraps or provincial rulers. The Bible says that Darius did this "so that the king might not suffer loss" (Daniel 6:2b). Daniel was one of these three ministers.

The title given to Daniel in Hebrew is *serâk*, which is translated in the Septuagint as *taktikós*, meaning the person in charge of running the operations of the kingdom and safeguarding the king's interests. This word is translated into English as administrator, governor, and supervisor. The tactical activities fall upon a group of individuals responsible for the organization and direction of any given enterprise. Do you want to be like the prophet Daniel? Lead the tactical actions of your ministry.

Strategy defines the tactical course, and tactics define the practical approach.

The tactical aspect of ministerial technology aims to run the operations of the kingdom of God and safeguard the interests of the King.

In the tactical order of the kingdom of God, the King takes preeminence. This is confirmed by Paul in his first letter to the Corinthians with reference to the resurrection:

> But each in turn [*tásso*]: Christ, the firstfruits; then, when he comes, those who belong to him (1 Corinthians 15:23).

In the order of things: First Christ, then us. The resurrection of Jesus Christ was God's tactical operation for His strategic purpose of salvation for all humanity.

When Jesus Christ gave the primary instruction for the work of the church, He commanded the disciples to wait for Him at the designated place (Matthew 28:16). The same verb *tásso* appears here in a passive form, this time referring to what they had agreed upon.

Tactics involve mutual agreement. They specify the commitments made and indicate the actions to be taken. As we will see later, the exercise of the mission is the perfect fit for all the tactics of technology in the ministry of the church.

Tactical Factors

As tactical operations are situational, actions are subject to constant changes and adaptations. A good tactic should consider various factors when making decisions and establishing routine systems to achieve goals.

Some of these factors to consider include:

- *Geographical location and weather conditions:* Holding a concert in a park during the rainy season may not be a good tactic. Adjust your activities based on your geographical location and weather conditions.

- *Local infrastructure capacity:* Consider reliable access to public services such as clean water, electricity, transportation, internet, and communications as crucial. Ensure a stable and dependable infrastructure. Cloud infrastructure services excel in this regard.

- *Environmental hazards and other threats:* Staying to hold a worship service during a hurricane and disregarding official evacuation calls is a high-risk tactic. Holding in-person worship services during a pandemic could be considered irresponsible. Include the role of technology in your contingency plans, such as how to respond to emergency situations and how to recover in the event of a disaster.

- *Promotion and communications:* Address the public opinion in your awareness campaign and in your methods of communicating the message. If the church wants to implement technology, it should discuss technology and educate its users on its use.

- *Sympathies and reactions of the community:* Perhaps there was a time when the loudspeaker on the street corner worked, but from what I can remember, this tactic is quite annoying and ineffective. We need to learn to read the reactions of the listeners to speak to them in the language and medium they are willing to engage with. Launch a promotional campaign that is appropriate for your audience.

- *Total economy:* Everything comes at a cost, and technology is no exception. If it does not cost you money, it will cost you time for adaptation and learning the necessary skills to sustain your initiatives. Take the time to study the total cost of your projects, including the three T's: time, treasure, and talent.

- *Legal system:* As we saw in **Chapter 3**, policies are one of the five pillars that support technology, and this includes the local legal framework where you carry out your ministry. Typically, the public department for technology or transportation and communications regulates the use of technology in the nation. It is important to be familiar with your legal environment.

- *Other organizations:* The church is a community-based organism, which means it is accountable to supervisory organizations such as ministerial councils, boards of directors, or denominations, and it also leads subordinate groups or associations. You need to consider how your technology tactics will impact both upwards and downwards.

- *Local government:* If your ministry aims to create impact in society, it should seek the support of local governments. Look for opportunities to leverage the resources available from the government and collaborate on community assistance projects. Consider both the district and city government, as well as the state and even federal government. I understand that navigating this path can be delicate, sometimes with ethical implications, but that does not mean you should not consider it.

- *Strategic partnerships:* Strategic alliances and partners can help you gather resources that you may not have initially. For example, they can assist you in reaching your target audience more easily in exchange for your expertise or provide you with free software licenses in return for promotion. Make a list of desirable allies and actively pursue those partnerships.

- *Change management:* As the Greek philosopher Heraclitus of Ephesus said, "The only constant in life is change." [51] It surprises me that most of our ministries do not have a clear change management process. Random change can cause destruction, but controlled change can lead to innovation. This is an even more critical issue when implementing technology. Establish clear procedures to accommodate the ever-changing nature of technology.

My intention in providing you with a list like this is for you to bring each of these tactical factors to the discussion table or ministry board so that informed decisions can be made, anchored in your local realities.

Operational Matters

Consider the following regarding operational matters. Each of the tactical factors will lead to concrete actions to be resolved in a timely manner. For each action, you must define the what, who or whom, and when, with an expiration date to ensure that the agreed-upon tasks are completed.

As an example, let us say your church wants to live-stream the worship service every week, and you want to use the tactical factors to define your operational actions. For the sake of space, let us consider only two of them: infrastructure and change management. I suggest you write something like this:

Infrastructure: We do not have servers open to the internet, so we need to use platforms provided by public companies. We have decided to use a social media platform for live streaming.

- Action #1: John will create a follower's page on the platform for next week. Alicia will assist him in creating the corporate identity elements. The pastor will announce and promote the page in two weeks for the launch during the weekend services.

- • Action #2: Peter will download the software to control the camera shots for next week. Alicia will assist him in creating the corporate identity elements to be incorporated in the video, such as logos, colors, and overlay titles. The pastor will conduct a test next Saturday where Peter controls the transmission and John connects it to the live stream as a private broadcast.

As you can see in the previous example, the tactical factor leads to specific operational actions in time. Once the program is underway, these actions will become recurring at the determined frequency.

Change management: Our challenge is to be able to respond to the changing interests of online believers over time. We have decided on the following actions to address the changes in interests.

- Action #1: The pastor will use social media comments to create a survey that helps capture the needs of believers. Peter will program the survey using specialized software to collect data quarterly. Peter will promote the survey through social media when the time comes.

- Action #2: Peter will regularly update the streaming software following the technical support schedule. Peter will provide the collected data to the pastor and the leadership team. The pastor will analyze the data to adjust the content, format, and sermon topics. Alicia will adjust the identity elements based on the suggested changes.

The above is just a guide on how you could document the problem, decisions, and actions resulting from those tactical decisions. You can use any format that you find most suitable for your way of working.

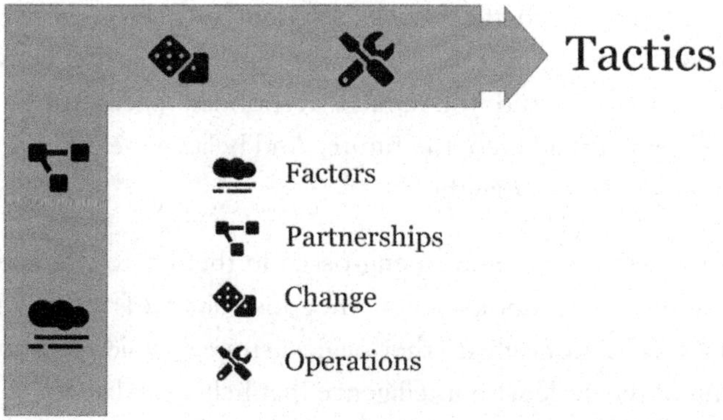

Figure 14: Tactical Elements

In the previous section, I discussed the importance of writing your strategy. Similarly, your tactical actions should be documented in an easily accessible document that you can incorporate into regular discussions with your team. This way, you can track your progress and determine the next steps to take.

Intelligence

Back in the Matrix, Neo arrived at a deserted square where an elderly woman was waiting for him, sitting on a bench, feeding crows. Neo approached cautiously and walked around the bench. As the saying goes, "Beware of those who surround themselves with crows..." It was her, the Oracle. She invited him to sit beside her for a conversation.

During their enigmatic conversation, which raised more questions than answers, Neo came to certain conclusions that would determine the future of humanity and their liberation from the Matrix. The Oracle was part of the system, a program that could scrutinize all the details of the Matrix and had the ability to collaborate with humans to plan a future of mutual benefit.

Neo said to her: "I suppose the most obvious question is, how can I trust you?" To which the Oracle responded "Bingo… you either accept what I'm going to tell you or reject it." Neo asked her again: But why help us?" And the Oracle responded back firmly: "I'm interested in one thing, Neo, the future. And believe me, I know – the only way to get there is together."

Many ministries become so engrossed in the day-to-day activities and operations of technology that they lose sight of the future, or rather, the vision. Not only do they neglect strategic guidance, but also the closing of the cycle with intelligence that helps validate the current state and progress towards their strategies.

Like the Oracle, intelligence should be part of your system, driving you to gather all the details concerning your strategic matrix and transporting you to the future you have already envisioned.

Intelligence is related to success metrics and the study of trends. It provides a detailed reading of all relevant indicators so that you can make calculated adjustments and continue moving steadfastly towards your future vision. This will help you make informed judgments and affirm your confidence in crucial decision-making.

Intelligence relies on two pillars: the collection of reliable data and the analysis of that data for decision-making. It is the feedback process to determine whether you are on the right track. Intelligence turns data into actionable information. It provides you with what you need to know to adjust strategies and the course of tactical operations.

Intelligence is critical and self-critical. It asks tough questions and perseveres to find truthful answers. Without answers grounded in harsh reality, we could lose our way and end up in an undesired destination.

Business Intelligence

Information Technology offers a whole discipline dedicated to what is known as Business Intelligence (BI). Business Intelligence combines data warehousing, data mining, and data visualization with business system analysis in aggregated tools to facilitate decision-making.

There are specialized software programs for Business Intelligence that connect to various data sources to create that aggregated effect. They often combine some expression of the following concepts for data presentation to the user:

- *Databases*: They use a structured data storage model, which means that data is stored in tables that are related to each other. We interact with them using Structured Query Language (SQL).

- *Data Warehouses*: This is a type of specialized database for extracting, transforming, and loading data from structured databases, performing calculations in the process, and combining relevant information for the business or ministry.

- *Object Stores*: They use an object storage model defined in a programming language according to the convenience of the software application being developed. For example, an object can be programmed to store attendance statistics in a text file every week.

- *Hierarchical Datastores*: It uses an unstructured model to store information in predetermined hierarchies. A good example of this form of information storage is the file system model on your computer's hard drive. There are document databases for this type of information.

- *Data Lakes*: It combines multiple structured, unstructured, object, and document sources into a broad collection space to organize and process large volumes of data in advance, before it is required for analysis.

- *Big Data*: As the name suggests, these are large-capacity data stores that have been optimized to store, retrieve, read, and manipulate a lot of data from any of the sources described above at the same time.

- *Data Mining*: It is the process of extracting data to discover repetitive patterns and relevant statistics that are not easily visible in a data lake. It may combine elements of artificial intelligence and a discipline known as Machine Learning (ML).

- *Data Visualization*: There are specialized applications that take aggregated data and present it graphically in a variety of formats and colors. Examples of these visualizations include alerts and notifications, trend charts, bar or pie charts, information cards, maps, and organized tables on a central dashboard.

I am convinced that the ministry can benefit from a Business Intelligence program that allows them to generate reports with visual aids and statistical analysis based on a collection of real and consistent data.

It is likely that your ministry may not have the finances or expertise to implement specialized software, but if you sow the concepts of Business Intelligence as soon as possible, you will be able to reap its benefits.

To start, you can use spreadsheets with clear metrics to collect key data frequently. The same spreadsheet has simple-to-use data visualization elements that can guide you in making wise decisions.

Now that we know where to store the data, the most important question is: What data should we store? How do we determine what things are important to consider over time, and how will we review it to inform our decision-making?

Key Performance Indicators

I was very young when I noticed that ushers used to walk along the church aisles during the sermon. I did not know why they did it, but my curiosity grew until one day I pretended to use the bathroom and stood to observe one of them. They had a small object in their hand that they would click. It was a counter. As they walked through the aisles, they counted the number of people sitting in the pews. This must be the most primitive form of business intelligence ever used in the ministry.

Churches often count two things: the number of people attending services and the money collected on the offering plates—the givers and the gifts. But there are many more things worth measuring to uncover the life of your ministry programs.

It starts with having clear objectives. Around those objectives, key questions are raised. These questions should give you an idea of the things you should measure. You can use questions like:

- Is it easy to measure?
- Does it measure the objective we set?
- Are we measuring progress, advancement, or regression?
- Can we change the results with our actions?
- How do we know if we achieved it?
- Does it tell us when it is time to reassess the objective?

The measurements you consider important are known in the world of business intelligence as Key Performance Indicators (KPIs).

Key performance indicators (KPIs) are established when critical data is collected from multiple sources and stored over the duration of the objective. This data is then organized and prepared for analysis with your ministry team. The organized data is presented in visual reports and helps uncover whether adjustments are needed in the strategic plan or ministry tactics, such as continuing or discontinuing a program or offering a new community service program.

You can visualize this process in **Figure 15** below.

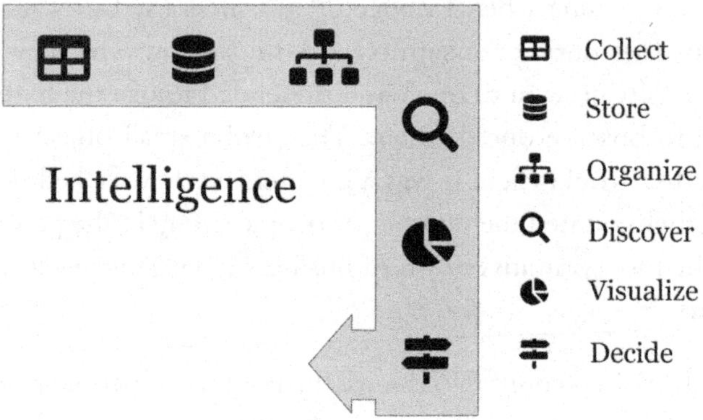

Figure 15: Intelligence Processes

Tracking the numbers of your ministry does not make you any less spiritual. In fact, it is so spiritual that there is even a book in the Bible named Numbers. Although it may seem contradictory, it is a critical activity. Numbers are important for informing our decisions. However, it is not about reducing ministry to numbers because we know it is much more than that. That is why it is essential to include both quantitative and qualitative factors.

Let me give you an example. If one of your strategic objectives is to disciple 50 people in a year, it is obvious that you need to count the number of disciples in the discipleship program; this is your quantitative measure. But you also need to measure the spiritual growth of each disciple and how you will assess it. You could guide them to keep a prayer and Bible reading journal and report their progress. This would be your qualitative measure. You want to measure both quantity and impact.

Therefore, in the strategic design process, for each strategic objective, you should assign at least one metric. This metric becomes your key performance indicator.

Your KPIs should measure beyond the obvious. Here is a list of possible metrics divided into three categories that you can consider in your strategic planning.

Attendance

It is not just about measuring overall attendance but being specific according to the programs your ministry offers. Consider the usage of your digital presence resources. Here are some attendance metrics to consider:

- *Attendance by membership status*: member, visitor, sympathizer, minor, contributor.
- *Attendance by age groups*: infants, toddlers, children, youth, young adults, adults, and seniors.
- *Attendance in common interest groups*: small groups, singles, married couples, college students, entrepreneurs.
- *Attendance in digital activities*: live streams, webinars and online communities, participation in social media groups, responses to email campaigns, comments and views on pre-recorded content.

Regarding digital activities, many social media platforms gather this information and provide dashboards to visualize the reach and engagement of your followers. Make sure to include these metrics in your planning and review meetings to inform your decisions.

Remember, these are just examples, and you should tailor the attendance metrics to fit the specific programs and initiatives of your ministry.

Tracking the numbers of your ministry does not make you any less spiritual.

You can also measure attendance and participation through the responses you receive using surveys. Survey applications, some of which are free, have their own success metrics. Surveys are an excellent way to capture both quantitative and qualitative information.

Volunteers

An active church embraces the priesthood of all believers, which means that everyone serves in some capacity according to the gifts received from the Holy Spirit.

That is why volunteering is a measure of the vitality and impact of your ministry in fulfilling its mission. We have already discussed the importance of aligning your performance metrics with your strategy, and this is particularly evident in volunteerism. Therefore, be as specific as possible when measuring the participation of your volunteers.

- *Volunteers by attendance*: Review the previous attendance metrics again and cross-reference each one with the number of volunteers who serve regularly. For example, volunteers per members, volunteers per elders, or volunteers per live streams.
- *Volunteers by mission area*: In the upcoming chapters, we will explore how to align your ministry's strategy with technology, and we will revisit this topic when relevant.

- *Volunteers by service program*: Make a list of the service programs offered by your ministry, if you have not done so already, and review the metrics of volunteer participation.
- *Digital and media volunteers*: Website, social media, email campaigns, graphic and visual presence, in-person production (sound, lights, stage, visual aids) for live streaming and broadcasting are some areas worth measuring.

You may have to deal with generational, demographic, and even cultural gaps. These may be key factors that are worth considering in your performance metrics as well.

You can also measure specific activities aimed at training, strengthening, and nurturing the lives and relationships of your volunteers. Remember that volunteerism is an opportunity to shepherd those who accompany you in ministry closely.

Volunteering is a measure of the vitality and impact of your ministry.

Do not forget to document and track these volunteerism metrics regularly to monitor the engagement and impact of your volunteers. It will provide valuable insights for optimizing your volunteer programs and fostering a thriving ministry community.

Decisions

There are decisions in the life of faith that can be excellent performance indicators. Do not overlook these. Decisions measure the direction in which your ministry is moving.

- *Sacramental decisions*: professions of faith, baptisms, weddings, healings and miracles, commissioning and sending of missionaries, among others.
- *Spiritual growth decisions*: measure participation in discipleship programs, preparation for baptism, ministry training center, symposiums on various topics, prayer activities, and fellowship.
- *Digital decisions*: social media trends include the number of followers, reactions, shares, comments, and video view retention.
- *Demographic changes*: while not direct decisions themselves, certain community life events determine their decisions. Do not forget to cross-reference data such as ages, deaths, nationalities, ethnicities, languages, and marital status with your performance indicators.

Money

It is obvious to measure the total money raised by your ministry. If it is a church, this often translates to the funds received in weekly activities. But there are many dimensions that can be measured when it comes to the finances of your ministry.

- *Regular income*: tithes, offerings, sales of items, tickets to seminars and concerts, registrations for camps and educational programs.
- *Income by service area or volunteering*: if your intention is, for example, for each service area to be self-sufficient in its finances, this should be measured.
- *Contributors by service area or volunteering*: another way to view money is by donor category. This considers factors such as generational or ministerial preferences of your donors.
- *Donations by type of payment*: cash, checks, credit cards, digital portals, cryptocurrencies, and securities, whether scheduled or sporadic, for the general fund or specific programs.

With so many things that can be measured in ministry, it is critical that you, along with your leaders or strategic planning team, choose which of these things are relevant to the strategic objectives you have set. Measuring these is essential for developing a good sense of how much progress you have made and how far you still have left to go.

And thus, we conclude the cycle of the exhilarating trilogy of technology (**Figure 11**). The well-known Austrian American management specialist Peter Drucker also coined the phrase, "The best way to predict the future is to create it." Strategy envisions the future you want to reach. Tactics put the wheels in motion to mobilize you. And intelligence informs the strategy to adjust the course and speed when necessary.

Now, let us see how these concepts apply to the mission.

MINISTERIAL TECHNOLOGIES

Mission is the mother of theology.

— DAVID J. BOSCH, AMERICAN THEOLOGIAN.

Missionaries make better products. [52]

— JEFF BEZOS, AMAZON.COM FOUNDER.

I had the opportunity to have lunch with Cardinal Paul Tighe, who at that time was the Secretary of the Pontifical Council for Social Communications for the Vatican and responsible for the new initiatives of the Roman Church for evangelization, including its foray into social media. Do you remember that Pope Francis was the first pope on Twitter @pontifex? Well, his first message was sent from Reverend Tighe's iPad.

During lunch, I asked him about how the Catholic Church was organizing the use of technology in ministry. I must confess that being the Catholic Church one of the churches with the most history and experience in these matters, I expected him to respond with a comprehensive, integrated, centralized, and strategic plan. But to my surprise and disappointment, that was not the case.

Despite knowing that the Church has been using some form of technology for centuries, that day he acknowledged that all initiatives were disconnected from each other, not aligned with a strategic plan, and did not even follow a common philosophy that defined the reasons or direction of technology in the Church.

The Catholic Church is not alone in this. This condition affects the entire Church, regardless of denomination or ecclesiastical tradition. While there may be exceptions, the absence of a clear model to address, organize, and even lead the use of technology in Christian organizations for ministry is the most common condition. In most cases, it does not mean that the Church is not using some form of technology to carry out certain areas of ministry, but rather that it lacks a holistic perspective and a strategic proposal that enables it to effectively incorporate technology in all areas.

This gap has given rise to *Cyberministry* as a model. Cyberministry is a source of information and resources to help Christian organizations comprehensively and collectively define the use of technology in their ministry.

My experience working as an IT professional and my leadership in ministry have allowed me a process of practice and reflection that has led me to the *Ministerial Technologies Matrix*® that I introduce in this chapter. My proposal is based on a simple premise:

The model for thinking, arranging, implementing, and leading the use of ministerial technologies is found in the mission of the Church. When we align the use of technology with the mission, and the mission, in turn, is aligned with the values and vision of the Church, we achieve strategic alignment of technology in ministry.

Let me break down this statement in some detail.

The Model

First and foremost, we need a model to follow. Many technological initiatives fail from the start because they do not follow a clear model. The lack of a model to define ministerial technologies has left the Church adrift and without organizational standards in this area. Imagine you are in the business of making chairs without having a model. How would you know if what you are producing are chairs and will function as chairs, and not astronaut suits? The astronaut suit might be very appealing, but it will not meet your need for chairs.

Similarly, when the Church dares to launch a technological initiative, how does it know if that initiative will yield the expected result? A model allows you to compare if the outcome aligns with the original intention.

Furthermore, the purpose of a model is not to limit creative capacity or the possibility of having a unique style of chair or ministry. The model should be flexible enough to adapt to the specific needs of the context. That is why having a model empowers and liberates the ministry to create something unique.

A model is indispensable if we want to foster understanding of ministerial technologies and positively impact our practice when employing them.

Until the time of writing this document, I have not found a clear model that allows the Church and other Christian organizations to fully understand their use of technology. Well, that is about to change.

TAIL

Surely, we have heard the expression "tail winds." The stern is the back part of any vessel (just as the bow is the front part). When sailing a sailboat with the wind at the stern or tail winds, two things happen: first, the wind does not create resistance to the vessel, and second, the wind exerts force against the mainsail. This allows the sailboat to move forward and reach its maximum speed, while the sailor can navigate with ease. Sailing with tail winds allows for maximum results with minimal effort.

That is the function of the model of ministerial technologies: maximum results with minimum effort.

That is why I like to use the analogy of tail winds as an acronym:

[T] - Think
[A] - Arrange
[I] - Implement
[L] - Lead

Think

Technological activity is an intelligent activity. By this, I mean that technology is not created randomly but is the result of the collective effort of thinking minds trying to solve real problems with appropriate tools.

The model of ministerial technologies allows you to think about technology from various perspectives, including a theological one. To get started on this topic, I recommend reading my article, *3 passages to formulate a biblical theology of technology.* [53] Any technological initiative for ministry purposes involves time for analysis and synthesis. There are thinking tools that can facilitate this process within your ministry teams, such as mind maps and decision matrices, for example. In **Chapter 11**, you will find references to these tools.

Arrange

Technological activity is an organized activity. Technology is a cultural way of arranging the material world. Any technological initiative for ministry purposes involves time for organization.

Order is a recurring theme in the Bible (1 Corinthians 14:40). Organization is the task of leaders. Moses organized the service in the sanctuary (Numbers 3-4). Saul organized the soldiers (1 Samuel 11:11). David organized the Levites (1 Chronicles 24:3) and the gatekeepers (1 Chronicles 26:12), and Josiah organized the great Passover celebration (2 Chronicles 35:16).

The model of ministerial technologies proposes organizing technology in a practical way around the five essential functions of the church's mission. We will explore the implications of this later. There are many things that require organization: people and their capabilities in the work of ministry, the systems they use, the devices they use, time, and money, to name a few.

Implement

Technological activity is a planned activity and implementation is the goal of any plans. There are many areas of technology that require planning. The first one is defining the role of technology in the ministry strategy, or, in other words, how the strategic plan obtains technological support, as we saw in the previous chapter. Then we have project planning, which allows for the full ideation to implementation of technology use.

There is a lot of planning in the Bible, so much so that it would be the subject of a complete work. Sometimes, we see grand plans to create monumental works. Noah's Ark was a great project. Similarly, the tabernacle in the wilderness and Solomon's temple. The conquest of the promised land by Joshua required a well-coordinated military movement. The gospel itself is the culmination of the millennium-long plan of salvation.

The model of ministerial technologies acknowledges that any technological initiative for ministry purposes requires time for planning. Familiarize yourself with the concepts and practices of project management in general, and technology management in particular; this will add value to your technological initiatives.

Lead

Technological activity is a managed activity. Moreover, management is a highly theological activity. Management is the first chronological commandment. Before God gave the Ten

Commandments to Moses, "The Lord God took the man and put him in the Garden of Eden to work it and take care of it." (Genesis 2:15). It is important to note that God, the Creator of all things, did not create the Garden of Eden but planted it (Genesis 2:8). The idea is that God is the first farmer and manager. He then delegated those functions to Adam in a process of knowledge transfer.

The word "administer" comes from Latin, combining *ad* (which means to add or sum) and *ministrare* (which means to serve or exercise a vocation), the same root as the word "ministry." The word "ministry" also contains the Latin word *minus*, meaning lesser, reinforcing the idea of humble service.

On one occasion, the disciples argued about who would be the greatest, and the Lord replied, "But you are not to be like that. Instead, the greatest among you should be like the youngest, and the one who rules like the one who serves" (Luke 22:26). Jesus revealed this hidden truth in Scripture: stewardship begins with servanthood (Genesis 25:23, 39:4; 1 Chronicles 5:2; Matthew 24:47; Mark 9:33-37; Romans 9:12; Philippians 2:5-11; 1 Corinthians 15:9-10). The steward is the servant of the house, so managing is the service that the church provides to those in the family of faith.

Stewardship begins with servanthood.

Every pastor or leader has seen in the exercise of their ministry the importance of being a good administrator, either through their own talents or by surrounding themselves with people who have that gift. This should not be done with arrogance, pride, or conceit. Nor by looking down on others. The role of the pastor is to care instead of being self-centered, to shield instead of protecting oneself, to elevate instead of elevating oneself, and to promote instead of self-promoting.

The pastor or leader thus becomes the facilitator for the people responsible for technology in the ministry.

In fact, administration appears as one of the gifts of the Holy Spirit. We read about this in 1 Corinthians 12:28:

> And God has placed in the church first of all apostles, second prophets, third teachers, then miracles, then gifts of healing, of helping, of *guidance*, and of different kinds of tongues (added emphasis).

In this passage, the word 'guidance' is translated from the Greek word *kybérnesis*, which is the same word that gives origin to the English cybernetics [54] and it points to the fact that administration is linked to the efficient and guided use of all resources: human, intellectual, financial, and material, all of which are necessary for the success of the church and its technological solutions.

Administrators or those who guide (*kybernétes*) are appointed by God in the church to think, arrange, implement and, of course, lead.

It is sad to see that most seminaries and Bible institutes that train our pastors and leaders do not provide them with the key tools for managing their respective ministry organizations. They do not train administrators with the same commitment and dedication as they train preachers and pastors.

The model of ministerial technologies implies administration applied to technology as an integrated function of personnel, systemic, procedural, political, and material resources.

As you can see, navigating the seas of technology is hard work. We cannot leave the ship adrift. Technology is an activity that requires commitment and effort. Every technological initiative with ministry purposes demands administrative time. What motivates us to use it in ministry is the impact it can have on the results.

Thinking, arranging, implementing, and leading are the TAIL winds that will carry your ship to its destination, measuring the results to know if you are achieving the desired goal. The model of ministerial technologies will help you start, grow, and account for your use of technology.

Keep these concepts in perspective as we dedicate a few lines to reviewing the mission of the church.

Mission

The church exists for its mission. Unlike other organizations that were born out of a particular activity or profession and later created a mission, the church was born for its mission. The mission gave it its initial origin and nature, and it should continue to provide the church with its purpose and ongoing identity.

The church exists for its mission.

For the church, mission is not just a statement, but an imperative. The early followers of Jesus Christ received this mission firsthand as a command in the Great Commission.

The earliest manifestation of this mission statement can be found in Mark 16:15, where Jesus said, "Go into all the world and preach the gospel to all creation." It is further expressed in Matthew 28:19-20a: "Therefore go and make disciples of all nations, baptizing them in the name of the Father and of the Son and of the Holy Spirit, and teaching them to obey everything I have commanded you."

Jesus' imperative is a call to mobilization for preaching and discipleship through baptism and instruction. Thus, the work does not end with evangelization but continues with the incorporation of the evangelized into the family of faith and their lifelong training.

As we will see next, several authors agree that the mission of the church consists of five essential purposes or functions.

Missiologist Charles Van Engen has devoted his life to studying the mission of the church. In a book that I consider fundamental for understanding the role of the local church as the recipient of Christ's commission, titled *God's Missionary People*, [55] Van Engen begins by showing how the concept of the church in the epistle to the Ephesians became a statement of faith that defined the church as "one, holy, and catholic" (understood here as the universality of the church, the church throughout the world). This model was later expanded during the Council of Constantinople, adding the word "apostolic" to emphasize adherence to the teaching of the apostles.

For centuries, the church defined itself as one, holy, catholic, and apostolic. Van Engen argues that these characteristics of the church should also define the mission of every local congregation, which is possible if we turn them into verbs instead of adjectives: to unify, to sanctify, to grow, and to proclaim. By doing this, the church begins to define its mission through:

- Communion and fellowship (*koinonía*) for unity.
- Testimony and learning (*marturía*) for sanctification.
- Service and assistance to others (*diaconía*) for growth.
- Preaching and outreach (*kerygma*) for evangelization.

From this analysis, it is evident that these four Greek words are key to defining the church on its mission.

Several years later, pastor Rick Warren, in his influential book *The Purpose-Driven Church*, [56] added a new (and necessary) dimension to the mission by joining the Great Commandment with the Great Commission. Warren said: "A Great Commitment to the Great

Commandment and the Great Commission will grow a Great Church."

The great commandment, of course, is found in Matthew 22:37-40, which reads: "Love the Lord your God with all your heart and with all your soul and with all your mind... [and] Love your neighbor as yourself. All the Law and the Prophets hang on these two commandments." The Great Commission, as mentioned before, is found in Matthew 28:19-20a.

When these two are combined, they define what Warren calls the five purposes of the church, which are: evangelism (go and preach), discipleship (make disciples), fellowship (baptize), worship (love God), and ministry (love your neighbor). Warren agrees with Van Engen on his four purposes but adds a fifth one.

In a personal conversation with Van Engen, he acknowledged that his model lacked that fifth component: worship or liturgy for the church to express its love for God. This theme, by the way, also resonates in the book of Ephesians, as we were created "for the praise of his glory" (Ephesians 1:6, 12, 14).

Church consultant Aubrey Malphurs, in his book *A New Kind of Church*, [57] prefers not to call these five components or purposes but rather functions, writing: "I prefer not to use the term *purpose* because I believe that the church's single purpose biblically and theologically is to glorify God" (let us remember Ephesians 1). However, he agrees that the mission of the church has five essential functions, which are "Teaching, Fellowship, Worship, Evangelism, and Service."

Another author who agrees with Malphurs that there are five essential functions of the church's mission is Alan Hirsch. In his book *5Q*, he refers to these functions as the intelligence (hence the title) and the original capacity of the body of Christ.

Hirsch arrives at a similar conclusion by using Ephesians 4, particularly verses 11-13, which read:

> So Christ himself gave the apostles, the prophets, the evangelists, the pastors and teachers, to equip his people for works of service, so that the body of Christ may be built up until we all reach unity in the faith and in the knowledge of the Son of God and become mature, attaining to the whole measure of the fullness of Christ.

Hence, the five intelligences (5Q) of the church: being sent as the primary agent of mission in the world (apostles), attending to the things that matter to God (prophets), sharing the story of the Gospel to invite people into a relationship with God (evangelists), nurturing and developing a healthy relationship with the faith community (pastors), and enriching disciples with knowledge, wisdom, and understanding (teachers).

These five intelligences encompass different aspects of the church's mission and highlight the diverse roles and functions that contribute to the overall purpose and ministry of the church. [58]

When only some of these functions are performed, or when priority is given to certain functions while neglecting others, the church becomes imbalanced and fails to grow into the maturity it is called to. In the words of Hirsh himself:

> We become incoherent…dysfunctional. We cannot be the body of Christ as Jesus intended without the fivefold active and present throughout the life of his movement. Said in another way, the church's capacity to 'attain the fullness of Christ' is at stake. [59]

The table in **Figure 16** below shows the comparison of different authors regarding the mission; I have rearranged each author's list to align it with the one I provide.

I prefer Van Engen's idea of using verbs instead of nouns since the mission is what the church needs to keep doing to achieve its vision – to understand the relationship between mission and vision from a strategic perspective, be sure to read the previous chapter on the trilogy of technology.

Van Engen	Van Engen (corregido)	Warren	Malphurs	Hirsh	Lugo
	(liturgia)	Worship	Worship	Prophetic	Adore
Sanctify	*marturia*	Discipleship	Teaching	Teaching	Assimiliate
Proclaim	*kerygma*	Evangelism	Evangelism	Evangelistic	Announce
Grow	*diakonía*	Ministry	Service	Shepherding	Assist
Unify	*koinonía*	Fellowship	Fellowship	Apostolic	Associate

Figure 16: Mission Essential Functions Compared

In summary, we can look at the mission of the church as the integration of these five essential functions: Adore, Assimilate, Announce, Assist, and Associate.

A healthy church balances these five actions (5 "A's") of mission (**Figure 16**). A healthy church is not neglectful of any of the five functions, but rather does everything – all its programs and events, all its interactions with the community and the world – under these five actions. A balanced church reflects in its finances that it attends to the five actions. A healthy church aligns all its technology with these five actions.

Let us break down this model in some detail to understand some of its practical aspects.

Adore

The first commandment in priority. Derived from the great commandment to love God (the first love), it finds its corporate expression through liturgy. Before you judge me for using this word, let me explain that the word liturgy comes from the Greek verb *leiturgéo*, which means to serve publicly, and it refers to the order and form that the ceremonies of public worship to God take.

Although worship is a personal responsibility, we cannot underestimate worship as the corporate action of the body of Christ. In fact, the same can be said for each of the essential functions; each essential function has a personal and a corporate component.

Assimilate

Jesus' commission was not just to preach the gospel but to make disciples. Discipleship involves an ongoing relationship for the rest of the believer's life. Learning includes all activities related to being a disciple and discipling others.

Training to live the Christian life in holiness, instruction, and teaching from the apostles in the Scriptures, and the reproduction of the mission are all elements of discipleship.

The responsibility of the church is to ensure the growth of believers towards maturity, accompanying everyone with instructional programs, preaching, educational institutions, and everything necessary to help them develop the character of Christ in faithfulness and love and to become disciples and disciple makers.

Announce

Jesus told the disciples before His ascension, "you will be my witnesses in Jerusalem, and in all Judea and Samaria, and to the ends of the earth" (Acts 1:8), which echoes the mandate for evangelism in Mark 16:15. The proclamation of the gospel is a dynamic action, always on the move, always conquering new frontiers, new digital and physical worlds.

In the Western Christian world, we sometimes have the false impression that we have already reached all the peoples of the earth, but nothing could be further from the truth. A little research can reveal that there are unreached groups even within the borders of countries traditionally considered Christian.

This is also true in the virtual realm. Each social network can be seen as a digital nation. For example, video games have created digital worlds that have not been reached by the gospel. The metaverse should challenge us not only to be the universal church but also the metaversal church. [60] The imperative is still valid.

> *The metaverse should challenge us not only to be the universal church but also the metaversal church.*

Assist

The second commandment in priority. The exercise of loving our neighbor as ourselves (Mark 12:31). Note that I refer to love as an exercise, meaning an action. In our culture, there is much talk about love as a romantic relationship, but in the Bible, love is something we do.

Jesus never said "I love you" to anyone, but He showed His love by giving His life to rescue us. Love for our neighbor should find expression in serving others, ministry, pastoral care, and various forms of assistance to others, ranging from community impact and public health to social justice programs.

Associate

All the previous functions depend on this one. We do not worship in an individualistic and isolated manner, as I mentioned before. We also cannot learn from others or help others if we are not in a mutual relationship. And of course, why reach out to others if it is not for us to walk in faith, worshiping, learning, helping, in short, accompanying one another?

In the New Testament, the phrase "one another" (*allos pros allon*) appears sixty-two times. Some verses invite us to love, care for, bear with, wait for, build up, and even wash one another's feet. Others confront us not to hate, betray, lie to, attack, or bite one another.

This implies that the mission of the church is exercised, expressed, maximized, and fulfilled in the community of believers. In a world where individualism is rewarded and people are fascinated with the self, the function of accompanying one another is countercultural but indispensable.

To accompany one another in faith means to walk together in unity or *koinonía.* If we desire to accompany one another, we must answer the question posed in Amos 3:3: "Do two walk together unless they have agreed to do so?" The answer to this rhetorical question is obvious; we must walk together in agreement. Recognizing the elemental functions of the mission and having a clear strategic plan helps us come to an agreement to walk together.

Before we continue, let us review our premise: the model for thinking, organizing, planning, and managing the use of ministerial technologies is found in the five "As" of the church's mission (Figure 17). We can envision and create strategies, but the mission of the church has remained the same since its inception.

Figure 17: 5 As of Mission

Now we need to tie together the definition of technology that we worked on in **Chapter 3** to arrive at a new concept of what constitutes ministerial technologies.

Technology in Mission

As you have observed, I lean towards a practical definition of technology, and I also lean towards a practical definition of mission. Ministerial technology is technology in practical mission. The idea is that it finds expression not only in what the church is and does—or should be doing—but also in the day-to-day use of technology in ministry.

Note that technology, like mission, consists of five components that form an interconnected network dependent on culture, which establishes its context and range of influence. This means that any implementation of ministerial technology must consider communication patterns, intergenerational relationships, the so-called "digital divide," economic capabilities, ethical implications, social inequality, and the impact of future innovations.

Ministerial technology is technology in practical mission.

Based on your definition of technology, I propose the following definition for ministerial technologies:

> Ministerial technologies are an interconnected cultural network of people, systems, processes, policies, and devices that enable the church to fulfill the five essential functions of its mission in the world.

Ministerial technologies are practical tools that enable the church to fulfill its mission effectively and efficiently. They encompass a wide range of analog, digital, electronic, and informational resources used in ministry, including communication platforms, multimedia tools, management systems, online platforms, and other technological solutions. These technologies are strategically integrated and employed to support worship, discipleship, evangelism, community engagement, and other essential functions of the church. The aim of ministerial technologies is to enhance and facilitate the ministry efforts of the church, ultimately advancing its mission in the world, to the ends of the earth.

Each of the five components of technology serves each of the five essential functions of mission to think, arrange, implement, and lead the ministerial technology. This is represented by the matrix in **Figure 18**.

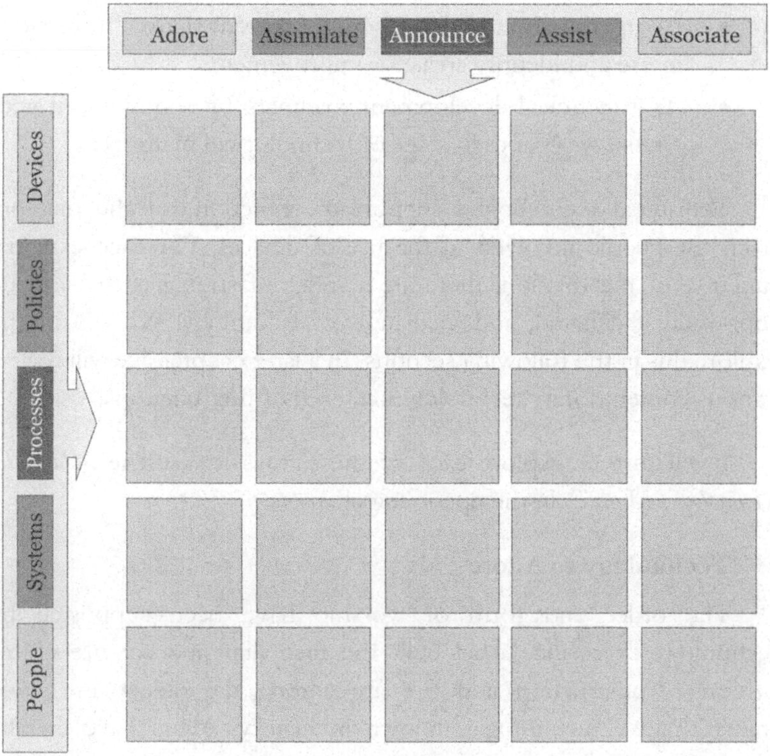

Figure 18: Ministerial Technologies Matrix®

The matrix is a practical tool. It is used to determine the role that each of the five components of technology will play in each of the essential functions of your ministry. Visit the *Cyberministry* website to obtain a printable copy and use it as a discussion guide with your team.

You can use the matrix in various ways, for example:

- As a working template with the TAIL model in each of the intersecting boxes, as explored in the rest of this book.
- As documentation of the technologies currently in use.
- As a planning guide for technologies that are being implemented.
- As an evaluation tool to assess the current state of technology in use and identify areas for improvement.
- As a technical development roadmap, by combining it with *Cybermaturity* at each stage of technological maturity.

Ministerial technologies support the functions of the mission, from the people involved to the use of devices. For each essential function of the mission, there are a series of strategies that can be supported, facilitated, and mediated by technology. We will briefly explore this in the following sections. In a later chapter, we will review various contemporary technologies in each of the functions.

I will further explore each of these ways to use the *Ministerial Technologies Matrix®* in the upcoming chapters.

Technology to Adore

The order and form of worship have been supported by technology ever since Abel built the first altar and set fire to his offering. Construction and fire are among the oldest and most transformative technologies known in history. Altars have literally borne the weight of worship offerings throughout the ages.

Biblical worship is multimedia and multisensory thanks to the support of various technologies. The priests of the Old Testament, for example, maintained the sacrificial system with its processes, regulations, and all the altar utensils used to present pleasing offerings to God. The same can be said of the Levite singers, with their training schools, rotation system, and the use of all kinds of musical instruments. Even the act of writing to preserve their poems and songs was mediated by the technology available to them.

Biblical worship is multimedia and multisensory.

Similarly, many technologies support or enhance the worship experience in contemporary churches. The lighting of sanctuaries, sound systems, acoustic and electronic musical instruments, personal and family devotionals, various formats of Scriptures, diverse visual aids, such as incense and ointments, and even the communion or Lord's Supper, allow our worship experience to engage all our senses (bread and wine being products of specific production techniques).

All these devices require a combination of people, systems, processes, and policies to function fully and effectively.

Technology to Assimilate

Among other things, the Bible is a manual for life. Its pages are filled with teachings and instructions, and any church learning initiative should be grounded in it. The fundamental piece of technology for learning is, therefore, Scripture.

Patriarchs, prophets, scribes, Pharisees, kings, pastors, and teachers have created comprehensive systems for preserving, reproducing, and instructing to safeguard the memory of the Scriptures. They developed oral traditions and writings on stones, skins, and papyri, phylacteries, [61] and graphic representations in murals, stained glass windows, and illustrated books.

The Reformation of the Church was largely made possible by the rapid reproduction and dissemination of the Bible in the language of the people, thanks to the implementation of a new technology for its time, movable type printing press. With new translation techniques and mass reproduction of the Scriptures, discipleship reached a new level.

Similarly, the technological explosion of the twentieth century has resulted in access to new ways of preserving, communicating, translating, and teaching the Scriptures, and modeling life according to them. Distance learning through electronic means, mobile telephony, video conferencing, various software versions, and biblical libraries, and multimedia formats are some examples.

Technology has placed in the hands of ordinary people the ability to become authors and creators of fresh content through self-publishing, as part of what has been called the "democratization of media." I believe that, overall, the church is still in its infancy when it comes to exploring the possibilities of new technologies for discipleship.

Technology to Announce

The early Christians wholeheartedly obeyed Jesus Christ's imperative to go into all the world. The promise of the Holy Spirit before His ascension empowered them to be witnesses from Jerusalem to the ends of the earth (Acts 1:8). After the experience of Pentecost (Acts 2:2), they set out on their mission.

In less than a century, the message of the Gospel reached every corner of the Roman Empire thanks to the technology of that time. The Pax Romana allowed the early Christians to move freely to fulfill Jesus' mandate, utilizing the sophisticated system of roads that connected to the capital of the Empire, as well as maritime transportation, which enabled faster and safer travel than by land.

The message spread among the technologists of the time: the craftsmen or *technítes* who worked in producing material goods to finance the work of the Gospel, with the apostle Paul being one of them.

The existing network of synagogues in major urban centers facilitated the penetration of the Gospel. The Church also used a system of manual copying and rotation of the Gospels and apostolic letters to circulate them promptly among various congregations. In subsequent centuries, Christian writings of various kinds proliferated, including instructions for discipleship and apologetics to reach and convince antagonists.

Much more could be said about how various technologies have been used throughout history to reach unreached people with the Gospel. I mentioned, for example, the role of the printing press in the Protestant Reformation and its counterpart in the Catholic Counter-Reformation. The so-called discovery of America also extended the reach of Christ's message to other peoples of the world.

The twentieth century witnessed tremendous leaps in outreach capacity with the introduction of many communication technologies that built upon one another. Starting with the telegraph and telephone, moving on to radio and satellite and cable television, and finally to the Internet and social networks, the boundaries of the ends of the earth suddenly shrank.

Similarly, in transportation, the arrival of steamships, trains, automobiles, and airplanes, opened immense opportunities to reach a world in need of hearing the Gospel.

The current problem for many Christian ministries is not access to outreach media, but the lack of strategy, reflected in the poverty and mediocrity of their programs. If we want to reach the new generations, we must reconsider how we use these media to convey a relevant message that does justice to the medium being used. Let us move away from the pulpit mentality[62] dominating our media.

Let us move away from the pulpit mentality dominating our media.

Technology to Assist

The Bible is full of provisions to help others, particularly those in conditions of extreme need. It often speaks of caring for the Levite, the orphan, the widow, the persecuted, and the foreigner (Deuteronomy 10:18, 16:11-14, 24:17-21; Psalms 146:9; or Jeremiah 22:3).

The agricultural system was meant to be generous, leaving enough in the fields for the needy. Psalm 68:5 declares, "A father to the fatherless, a defender of widows, is God in his holy dwelling." In other words, the house of God or the temple was also meant to function as a place of refuge and public assistance. The application of justice was not to favor the rich and powerful.

You might be wondering, what does this have to do with technology? Well, a lot, when we consider that agriculture is technology at its fullest, that the temple is material culture for worship, and that justice is a system. All these things are technologies, even if some of them do not involve a specific device. Judges, elders, and counselors relied on the legal system with its procedures to execute written laws and document their verdicts.

There are assistive technologies that help individuals with disabilities or learning difficulties. With accessibility technologies, digital resources are made available to a larger number of people. The mobility of digital money allows remittances to reach individuals with limited resources, churches in crisis situations, or even foreign missionaries.

We should explore the role that the Internet of Things (IoT) can play in providing services to those in need, including essential items, healthcare services, and shelter, as well as creative ways to help individuals with their finances, addictions, and interpersonal relationships. Using technology to transport provisions from a food bank to meet the needs of the community is also a form of technology.

Helping through education is another way, providing vocational and professional guidance courses and workshops in computer labs, and providing training for the elderly to benefit from the latest advances in personal communications.

All these examples demonstrate the intersection between technology and the biblical call to assist those in need. Technology can be a powerful tool in meeting the physical, emotional, and spiritual needs of individuals and communities.

Technology to Associate

Associating encompasses all the tasks of the ministry, including management and coordination of those tasks. Associating involves accompanying and administering the organization to stimulate a sense of belonging, identity, and fellowship through mutual service and proper allocation of material resources for the benefit of those who make up the ministry and those reached by it.

The Bible is full of examples related to accompanying others. The work of constructing the tabernacle in the desert, the distribution of land during the conquest of Canaan, and the project of rebuilding Jerusalem under Nehemiah were all community activities that the people could not have accomplished without walking together.

Many technologies support the process of accompanying others in the ministry of the church by facilitating the management of human, intellectual, financial, and material resources. These include enterprise information systems, which, when applied to the church, are known as church management or membership systems. With such software, finances, personnel, payroll, participation, attendance, and volunteering can be managed from a central platform.

Internet technologies such as email, social media, and websites are crucial in the process of associating with others because communion relies on communication.

The church is a knowledge enterprise and content producing machine. Recorded sermons and teachings in audio, video, or written formats are useful tools for accompanying others on their faith journey. Content management and intellectual property protection systems can find expression in the church, given the wide range of content produced each week. In general, information technology offers modes of storage and preservation for all these resources.

Anything related to mobility, transportation, and logistics also falls into this category. For example, providing transportation for elderly individuals to participate in in-person activities or attend medical appointments.

As you can see, various ministerial technologies can overlap with each other, allowing a single strategy to cover multiple areas of mission.

This concludes our overview of ministerial technologies. The purpose of this bird's-eye view is to lay the foundation for the following chapters. I invite you to consider the potential impact certain technologies may have on your ministry and to experiment with them along the way as you employ them.

In the next five chapters, we will delve into each function of ministerial technologies in detail, with references to specific technologies and their utilization in the mission's work.

TECHNOLOGY TO ADORE

Make an altar of earth for me.

— GOD, EXODUS 20:24.

Church budgets are becoming larger, and worship technology is at the forefront.

MICHAEL FABAREZ – AMERICAN PASTOR.

I was tired and irritable. I had gone two consecutive sleepless nights, and it was the night of the third day. But seeing everything in its place, I knew it had been worth it. The next day would mark the beginning of the church's second Worship Symposium, and I had taken responsibility for decorating the sanctuary.

Inspired by the description of the desert tabernacle (Exodus 25-31), we recreated the Holy of Holies with columns, curtains, and representations of the Ark of the Covenant and the altar of incense. We made various processional elements such as ribbons, flags, banners, and even handkerchiefs that the entire congregation could use to participate in the celebration.

The event was an unforgettable and transformative experience for me and many others. I felt like a child presenting his latest drawing to his father. For the first time, I had created something with my own hands that would bring a smile to the face of my Heavenly Father. I had built an altar.

In the previous chapter, I explained that ministerial technologies are those that enable the church to fulfill its five essential functions: worship, outreach, learning, assistance, and accompaniment. From this perspective, there is technology for each of these five areas. In this chapter, we will examine in more detail the technology for worship.

Media Worship

I was engaged in an email conversation with a friend discussing an article about worship, and he wrote to me, "No one really knows what worship in the future will be like, but it is likely to be more technological."

He is right because the experience of worship has always been mediated by technology. It may seem strange to think of it in these terms, but it makes a lot of sense. The material medium is part of our expression of worship to God.

The oldest piece of worship technology is the altar. Worship has relied on the altar since Abel built the first one and set fire to his offering, and altars have supported the offerings of worship throughout biblical times.

Chapter 4 of Genesis tells the story of Cain and Abel. Each of them prepared an offering to worship God. Cain brought an offering of the fruits he cultivated, while Abel chose the firstborn and fattest of his sheep. Notice how the narrative describes Abel's offering as the firstborn and fattest, while it does not describe Cain's offering.

God much preferred Abel's offering over Cain's. When Cain learned of this, he became so angry that he took his brother's life.

Although the story is told in broad strokes, you can infer all the elements of technology. The offerings on an altar are part of the biblical cultural network. There is an interaction between people - Cain and Abel; in a system of interconnected parts – Cain's agriculture and Abel's sheep farming; a preparation process - it seems that Abel paid more attention to the process; a selection criterion or policy - the first and the best; and material elements - what they offered and on what they offered it.

These same elements will be present in other narratives, in other altars. You will identify them in the stone altars of the patriarchs, in the bronze altar of the desert tabernacle, and in Solomon's temple altar, which on the day of its dedication, could not fit all the firstborn fattened cattle offered by the people (1 Kings 8:64).

Biblical Worship

The Levite priests of the Old Testament were responsible for maintaining the constant system of sacrifices, with their processes, regulations, and all the utensils of the altar to present offerings of forgiveness, peace, and pleasing aroma. This dynamic was instituted by God Himself when the people of Israel were journeying through the desert.

The first mobile worship technology appears in the book of Exodus, chapters 25 to 40. I invite you to read the entire passage continuously. The details of constructing the tabernacle are extraordinary and even overwhelming. It had to be built meticulously and by the right people. God spoke to Moses and provided him with detailed instructions on how to build the tabernacle, including the dimensions, materials, and furnishings. God told Moses:

> Then have them make a sanctuary for me, and I will dwell among them. Make this tabernacle and all its furnishings exactly like the pattern I will show you (Exodus 25:8-9).

The exact replica of the heavenly detailed model included:

- The dimensions and architecture of the temple with its wings and divisions.
- The internal decorations with embroidered curtains and the overall external covering, in a way echoing the garden of Eden.
- The furniture in the courtyard: the altar for sacrifices and the bronze basin for the priests' cleansing.
- The furniture in the Holy Place: the table of showbread, the seven-branched lampstand, and the incense altar.
- The ark to store the stone tablets with the inscribed law, a budding almond rod, and a jar of manna, covered by the mercy seat crafted from pure gold with the wings of two cherubim covering it.
- The garments of the high priest and the assisting priests, including the ephod adorned with twelve gemstones representing the twelve tribes of Israel.
- All the vessels, utensils, knives, tongs, and other tools for handling the sacrifices.
- The boxes and cases to pack and transport everything to the next location.

I must confess that I find great delight in reading these passages from Exodus. They thought of everything, or rather, God thought of everything. I have been in churches that have expensive equipment for worship in the sanctuary (such as projectors, cameras, and microphones), but they forget to purchase cases to protect, transport, and securely store them. The tabernacle of worship came with instructions to store and move everything from one place to another.

For this magnificent work, highly skilled technologists were needed, with experience and expertise in a variety of technologies. Exodus 31:2-6 reads:

> See, I have chosen Bezalel son of Uri, the son of Hur,
> of the tribe of Judah, and I have filled him with the

> Spirit of God, with wisdom, with understanding, with knowledge and with all kinds of skills— to make artistic designs for work in gold, silver and bronze, to cut and set stones, to work in wood, and to engage in all kinds of crafts. Moreover, I have appointed Oholiab son of Ahisamak, of the tribe of Dan, to help him. Also I have given ability to all the skilled workers to make everything I have commanded you.

Much can be said about this passage. The first thing is that Moses hired a project manager who was also an expert in all the technologies that would be employed, as the director of the entire work, Bezalel, and his assistant, Aholiab. Remember that technology begins with people, not with devices.

By the way, Solomon also had a technician in charge of the temple work. Huram of Tyre, who was "filled with wisdom, with understanding and with knowledge to do all kinds of bronze work" (1 Kings 7:14). The rest of chapter 7 of 1 Kings describes everything he made.

Furthermore, it was a team effort. Moses would be responsible for collecting all the raw materials through a massive offering, while Bezalel and Aholiab would oversee the recruiting "all the skilled workers" who would carry out the manufacturing tasks. It is worth mentioning that Solomon had "two hundred and fifty officials supervising the men" (2 Chronicles 8:10).

The right people for the job are better equipped to define appropriate systems, productive processes, and sound policies to create and use the right devices.

Second, technology requires material resources to create new material resources. These material transformation processes are part of technology itself. In other words, you need technology to create technology. To set a precious stone in a piece of gold, you need mining to extract the gold from the earth and lapidary work to remove its commonness and make it appear precious.

> ***You need technology to create technology.***

Thirdly, the creative ability, intelligence, and wisdom of these men operated in the fullness of the Holy Spirit. The Holy Spirit can do the same today with the men and women in charge of technology for worship in your church or ministry. That is also my prayer. May your technological capacity be anointed by the Spirit of God so that you can do everything that your strategy guides you to do.

Lastly, various technologies come together to create something new. In the creation of the tabernacle and its parallel in Solomon's temple, there are many interconnected technical skills. In the table in **Figure 19**, I provide a list of the most obvious ones derived from the passages in Exodus.

Technologies found in the dessert tabernacle		
Metallurgy	Cabinetmaking	Sewing
Metalwork	Tannery	Instrumentation
Goldsmithing	Saddlery	Incense making
Jewelry	Leather working	Perfumery
Lapidary	Utensils making	Lightning
Stone setting	Weaving	Pyrotechnics
Carpentry	Spinning	Butchery

Figure 19: Technology to Adore

When you organize technology for worship in your church, be prepared for multiple technologies to interact with each other in your services and worship gatherings. We no longer need the butchery to be part of the worship —thanks to God for the sacrifice of Jesus Christ— but we do need other contemporary technologies.

Biblical worship technology extends to include the Levite singers, with their training schools, rotation system, requirements, and the use of various musical instruments, as expressed in the Bible on many occasions.

Interesting fact, the grandson of Cain, the fratricide, was the first recorded inventor of musical instruments in biblical history. His name was Jubal, and he devised the harp and flute (Genesis 4:21). Since then, musical instruments have been part of the worship experience of God's people throughout the ages. What a story of family redemption!

King David also invented and crafted musical instruments (Amos 6:5), played the harp (1 Samuel 16:23), and sang (2 Samuel 23:1). By the time Solomon was about to succeed David, there were four thousand musicians playing the instruments that David had ordered to be made (1 Chronicles 23:5).

Years later, during the dedication of Solomon's temple,

> The priests took their positions, as did the Levites with the Lord's musical instruments, which King David had made for praising the Lord and which were used when he gave thanks, saying, "His love endures forever." Opposite the Levites, the priests blew their trumpets, and all the Israelites were standing (2 Chronicles 7:6).

David had much to do with the organization of the royal worship team, which was also responsible for all the work of the tabernacle that would later be transferred to the temple. Kenneth Mathews tells us that:

> David established three musical guilds under his supervision. Asaph, Heman, and Jeduthun (Ethan) were Levites David appointed for the musical accompaniment of temple worship (see [1 Chronicles] 15:16–22). The guilds were drawn from all age groups and those with differing musical skills. They were divided into twenty-four courses like the priests. [63]

The Book of Psalms is a musical compilation of these ancient worship experiences. It is not possible to compare the music of that time with modern and contemporary musical advancements. In biblical times, there was no musical notation or electronic keyboards. However, the psalmists took it upon themselves to include musical notations to indicate how their songs should be performed. This system of musical writing and annotations or superscriptions is also technology.

> Many of the musical terms appear in the superscriptions, added by later scribes or collectors who sought to indicate the nature, authority, and use of psalms. These psalm headings list ancient categories of psalms, offer instruction for musical performance, refer to tunes, and identify occasions for a psalm's use in worship. Much of the information in these titles consists of specialized musical terms that readers no longer understand. [64]

I will not delve into a study of these musical notations, but they include *Selah*, which appears 71 times in the Psalms and is interpreted as a musical interlude, although its exact meaning is uncertain. Other words include *mizmor* (a song with accompaniment), *shir* (a song), *sakal* (instruction), and references to melodies lost in time, such as *shiggayon* (a lament), *tehilla* (praise), and *tephilla* (prayer). There are many more, but I believe these give you an idea.

Christians continued the Jewish worship practices. In the New Testament, there are no extensive texts about worship. However, it is well known that the disciples continued to attend the temple - the second temple in Jerusalem, rebuilt by Herod the Great, with all the objects of worship originating from the tabernacle built by Moses. Jesus himself had the custom of going to the temple to celebrate the annual feasts (Luke 2:46; Matthew 21:12-23; John 8:2; 10:23). The temple would become a platform for teaching and preaching about Jesus Christ (Acts 5:42).

Not only are we encouraged to sing and praise God (Ephesians 5:18-19; Colossians 3:16), but the New Testament also has its own "psalms" as there are recorded hymns believed to have been sung in the early congregations, such as the *Magnificat* (Luke 1:46-55), the *Benedictus* (Luke 1:68-79), and the *Nunc Dimittis* (Luke 2:29-32), as well as those found in the Pauline letters in Philippians 2:6-11, Colossians 1:15-20, and Ephesians 1:3-14. The Book of Revelation contains seven worship hymns that could allow us to develop a theology and praxis of Christian worship all by themselves (Revelation 4:8, 11; 5:9-13; 7:10-12; 11:15-18; 12:10-12; 15:3-4; and 19:1-2, 6-8).

Through the practice in the temple and in the gatherings of the early Christians, we know that worship technology was an integral element of the worship experience. This was reflected in architecture and art throughout the history of the church in different regions and cultures where it established local expressions.

Each local congregation should develop its own system of signs and notations to plan their participation and facilitate communication during live services. It is important to be able to curate, record, and store musical creations with an efficient and easily accessible filing system to be used when appropriate. Recording and playback of music are technologies for worship.

We are clear that worship is much more than technology. It is the exuberant expression of our love for God. It is a lifestyle that exalts the glory and beauty of God in everything we do. However, with this non-exhaustive overview that I have provided, it is evident that biblical worship is mediated and technological.

Understanding the biblical foundation of worship technology benefits you in several ways:

1. It generates faith and gives you permission, with a fresh perspective, to study the Word.
2. It justifies your technological programs and the budget expenses they entail.
3. It helps you plan the technical elements of your corporate worship with intentionality.
4. It allows you to define the systems and processes that require coordination and improvement.
5. It enables you to inventory and evaluate the devices you use for worship and how to improve the congregational experience.

Knowing these benefits, you can confidently create something with your hands that will bring a smile to the face of God.

Biblical worship is mediated and technological.

Adoring with Technology

Whether conscious or not, today's church also relies on and enhances the worship experience with various technologies. The church makes use of the architecture of its temples with all their functional requirements and urban standards; the lighting of its sanctuaries; the sound system and new acoustic and electronic musical instruments; personal and family devotionals and lectionaries alongside the many formats of Scripture; and various audiovisual aids, among other things.

All these technologies create a unique and complete communal worship experience for each local church. In other words, each local church must shape and purposefully use the technology to fulfill its goals.

I dare to think that this applies to any ministry or mission, parachurch organization, or educational institution. The act of worship should be embedded in its strategy, and the technology for worship should be part of its plan.

The task is to design a model that works well for your church, utilizing the talents you have, for the reality of the people who are part of your congregation, in their context and culture. Remember the "interconnected cultural network" of our definition.

When the COVID-19 pandemic began, many churches started streaming as a response to the closure of in-person gatherings. Pastor Antoine, upon learning that this was a quick way to share the worship experience remotely, announced to his congregation that they would start broadcasting on a well-known social media platform the following Sunday.

Sunday arrived with high expectations, but to his surprise, only two people logged in: his wife and the young person in charge of the livestream. And so it went for the following Sundays. As no one connected for three weeks, he took it upon himself to call each member of his congregation to ask why they were not joining the online service. To his surprise, the common response was, "Because we do not have computers or smartphones."

Antoine had to change his strategy and developed a plan for weekly phone calls to minister and bring the Word to each person. This kept his congregation connected and growing during this time of extreme difficulties.

I share this story to illustrate the fact that not all technology for worship is suitable for everyone. I repeat, each local congregation and ministry organization must define this according to its own context.

TAIL Winds

The intentional church, then, thinks, organizes, plans, and manages each of the five areas of technology for worship. In this section, I will mention different technologies as examples to help you in your own work.

I recommend that you do this as a local exercise but also involve your council, board of directors, or leadership group. I guarantee that in the wisdom of many counselors, you will find a plan that suits your needs.

In this section, I do not want to jump ahead and provide a list of providers and technology applications. I have reserved this for **Chapter 11**, where you will find a list of technologies and references to find recommended providers for each mission function.

For now, let us apply the TAIL model to each component of technology for worship.

Adoration People

Worship is personal, familial, and corporate. Therefore, as a ministry, we should think about providing tools for personal devotional time, as well as for worshiping as a family and as a congregation. At the same time, we need to provide appropriate systems to lead congregational worship during our gatherings.

Think: Who are your Bezalels and Aholiabs? Who are your Levites and psalmists? Who are your prophets? One key person must oversee worship. This person should have an assistant. Together, determine the worship leaders, producers, singers, musicians, sound technicians, lighting technicians, graphic or visual artists, community managers, and preachers. Do not forget to seek help from expert consultants in these areas.

Arrange: Define roles and responsibilities and establish a schedule of activities. In other words, determine who does what and when. Organize physical interaction with existing technologies. For example, decide who is responsible for setting up and dismantling cables or when to check the sound and lighting for each service. Organize rehearsals. Organize the sharing of graphic materials. Create a preaching schedule. There is much to organize. The goal is to ensure that everyone is on the same page.

Implement: In every organization, there are regular and special activities. Regular activities are recurring and ongoing, while special activities are unique events. This applies to weekly services as well as special worship events based on the church calendar (such as Easter and Christmas).

Plan the activities of your worship team. Also, plan for the future of technology that you may not currently have but would like to implement. Determine when you will engage in planning sessions, and when you do, make sure to document the plans.

Lead: Lead the entire team, starting with the worship director. Direct the organizational activities to review progress and results. Keep databases updated with contacts, activity schedules, participation confirmations, cancellations, progress of plans, and all available resources.

Outdated data and a calendar from the year 2000, except for historical purposes, are of no use. Data that is not kept up-to-date loses its usefulness. The same applies to the calendar. Continuously track your performance indicators. Determine who will be responsible for keeping these things current and the frequency with which they will do so.

Adoration Systems

Have you ever taken something apart and had leftover pieces? It has happened to me more than once. It is even worse when you lose a piece, and although things may work for a while, that missing piece turns out to be crucial for long-term functionality. When we consider all the pieces, we are acting systematically.

The worship tent was a system. The columns were joined by rings and hooks to the colorful curtains and exterior skins. Everything was stored and transported in special chests. Can you imagine if some of those rings were lost? In a system, even the smallest pieces are crucial. Another example of a worship system is the biblical calendar of seven annual feasts.

Review the concepts about technology systems in Chapter 3 of this book to review worship systems using the TAIL model.

Think: Technology comes in both large and small pieces. Take inventory of your current systems and document all their parts. Include, for example, the sound and lighting system, the rotation system for worship leaders, musicians, and preachers, the church calendar, the offering collection system, or live streaming.

When I say document, I mean creating a written document for each worship system. It is helpful to document all the details, including the brands and models of equipment along with their manufacturing and usage manuals. This will assist in knowledge transfer when new team members join your team.

Arrange: Answer the following questions: 1. What are the parts? 2. What is the rhythm? and 3. What other systems does it depend on?

For example, if you use a sound system, the parts may include microphones, cables, mixing consoles, monitors, and speakers. The rhythm could be a weekly cycle of setting up, testing, using, dismantling, and storing the equipment. It depends on purchasing equipment and having electricity available, including batteries. Each component of the system fits into an order that can be adjusted to ensure the system is ready to fulfill its purpose.

Technology comes in both large and small pieces.

Let us use an example to illustrate this. The sound technician arrives at the church, but the person with the keys to the equipment storage room and the individual responsible for turning on the electricity in the building have not arrived yet. Additionally, the secretary who was supposed to order batteries for the microphones forgot to do so. Organize the parts of the system to ensure that there are enough batteries, the first person to arrive at the church switches on the power, and the person with the keys arrives before the sound technician. Document the order of each system in the catalog documents.

Implement: The components, rhythm, and dependencies of the systems have an impact on the availability of people and their skill levels, as well as on the budget and resources available.

Plan the finances by creating a detailed budget for the short, medium, and long term. Take into consideration the time and talent requirements of your systems. Budget for the implementation of new initiatives and allocate funds for future projects. There are specialized budgeting software tools available, but at a basic level, a spreadsheet will suffice.

Lead: Control the parts, rhythm, and dependencies. Ensure that everything necessary is provided on time and with the right people involved. Extend the calendar and use cloud-based document storage for managing the systems.

Whenever changes are made to a system, update the calendar and documentation accordingly. Communicate any changes or adjustments clearly. Decide on a communication system to use, whether it is email, phone calls, or text messaging – treat communication as a system with its own parts, rhythm, and dependencies.

Adoration Processes

When considering processes, think about movement and transformation. The individuals involved in worship are moving the systems of worship to transform the worship experience.

In biblical times, the priests were responsible for various processes, such as crushing olives to produce oil for lighting and anointing, grinding spices to create incense, sacrificing animals, and preparing them for offerings, and the process of mobilizing the entire worship tent. Those who have worked in mobile ministries know the work involved.

Think: This is where you have a lot of work to do. As mentioned in the process section of **Chapter 3**, I recommend documenting these processes using flowcharts. Some key processes for worship include preparing the systems for the service, selecting and rehearsing songs, programming and producing the worship service (as shown in Example **Figure 7**), sermon preparation, and usher coordination.

Arrange: Processes are documented in flowcharts but are organized into a workflow. Ask yourself, "What tasks are needed to complete this process? Who will perform them? And what decisions need to be made?" Additionally, consider which processes can be automated and which can be delegated to gain efficiencies that may be hidden when processes are not documented and organized. Determine which processes require integration with others and note the touchpoints between them.

Implement: Plan workflows in a step-by-step manner. You can use questions such as "What would happen if a certain person does not complete this task by a specific date?" "What are the consequences of inaction?" "What other things would be affected?" By thinking in this way, you can identify areas where detailed planning is necessary for these tasks. Some project management software applications offer modules for managing not only project tasks but also workflow management.

Consider which processes can be automated and which can be delegated.

Lead: While maintaining healthy and well-coordinated processes, look for opportunities to improve worship processes. The continuous improvement process helps identify inefficiencies and apply incremental or innovative improvements. With your team, determine which worship processes are most critical, such as producing the liturgy or the weekly service, for example. Ensure you create clear procedure manuals for these processes. Maintain a repository of documents with flowcharts and the roles that individuals responsible for these processes should fulfill.

Adoration Policies

We often do not think about the policies surrounding technology, let alone worship and technology policies. However, they do exist, or at the very least, should exist. Policies are the rules, norms, and even laws that govern technology and its usage standards.

The broader body of laws in the Old Testament can be found in the book of Leviticus. As the name implies, this body of laws consisted not of moral laws, but ceremonial laws overseen by the Levitical priests. Its purpose was to standardize all activities related to worship in the desert tabernacle.

The ceremonial Levitical laws include:

- Laws concerning animal sacrifices such as livestock and birds, as well as produce from the land such as flour and oil.
- Laws concerning the consecration of priests, their purification rituals, their garments, and their worship activities.
- Laws concerning celebrations and special offerings that the people were required to offer annually.
- Laws concerning the relationship with God and others, including the care of the needy among the people.

For Christians, many of these laws were fulfilled and thus considered abolished in the ministry of Jesus Christ. However, it does not mean that we should ignore their implications for the Christian life. The demands of contemporary technologies require new policies that we must adopt, good practices that we should follow, and even laws that we must comply with.

Think: What are the policies that govern worshipers and worship technologists in your church? What are the best practices? The term "best practices" is quite common in the world of computing and engineering today. It refers to praxis policies, which are not legislated by authorities but are expected to be followed to harmonize with how others, in this case, other ministries, successfully employ them.

Write the policies for using technology in worship and keep these documents in an easily accessible place for the entire team. Turn them into official documents. Include rules for being part of worshipers, musicians, and singers; the job requirements for each role; appropriate handling of the equipment used; and best practices for participation and usage.

Arrange: Arrange the policies into categories. Some policies are for internal use, others for congregational use, and others for public information. Place the policies for the public consumption of worship technology in a visible location for everyone. For example, in the

United States, firefighters verify that temples and meeting halls adhere to established codes for occupancy capacity. I recently saw a sign outside a church that read:

> By entering this building, you consent to be photographed, filmed, and recorded, and you grant us permission to use this material in any media communication we deem relevant. Please do not enter the sanctuary with uncovered food and beverages.

Familiarize yourself with local laws. The pastor does not need to be an expert, but they should work with a lawyer who stays informed about the legal developments in the country and city.

Implement: Plan annual review sessions for your policies and the recruitment and selection activities for your contracted and volunteer team members. Plan for remodeling or adjustments that may be necessary to ensure your physical premises comply with local building codes and to protect your equipment from accidents and contingencies.

Review all documents related to warranties and equipment protection. Register your equipment to ensure it is covered by these warranties. Use the calendar to establish dates for policy and procedure reviews.

Lead: Policies are not unchangeable laws. Be flexible in adjusting them as your ministry changes, your congregation rotates and grows, your skills strengthen, your community brings new needs, and your cultural context evolves.

Keep your insurance policies up to date. Some policies are mandatory in certain locations. Fire insurance or coverage for accidents occurring near the temple are always a good idea. Manage the necessary legal funds to cover all your bases.

Adoration Devices

The most obvious aspect of technology. The tip of the iceberg. The visible and tangible. Among the devices we use in our temples and sanctuaries, we can see cables and microphones, batteries, various musical instruments, lights and consoles, cameras and mixers, screens, and projectors. Bibles in the back of the seats and the bulletin are also technologies for worship.

Proverbs 14:4, reads:

> Where there are no oxen, the manger is empty, but
> from the strength of an ox come abundant harvests.

In the same way, with the right devices and tools, the work of worship can be facilitated and enhanced. Worship devices play a significant role in creating an atmosphere conducive to worship and facilitating the various elements of the worship experience.

Make sure you have the appropriate tools for your worship to bear much fruit, exalt the name of God, and transform worshipers into His image.

Think: Tools come in two forms: hardware and software. Hardware refers to tangible devices you can touch, while software refers to intangible programs and applications. So, what hardware devices and software are you currently using? What are the needs of your church or ministry? How do these tools align with those needs and the strategic plan of your church or ministry?

It is important to have a clear understanding of what you want to achieve, what resources you have available, and what you may still need. Each ministry context is different, so it is essential to research what tools are applicable to your specific context and how you can adapt them to work effectively and fulfill their purpose.

Tools come in two forms: hardware and software.

Arrange: There is a lot to organize. Hardware devices often come with usage recommendations that should be followed. You may have noticed that even operating temperatures and storage environments are suggested in the manuals. Arrange the security of the equipment and ensure secure access to the software.

Designate a dedicated space to store the equipment and avoid moving it around arbitrarily. Assign responsible individuals for the care of the devices. The same applies to the applications you use. Organize the manufacturer's manuals, warranties, and your own usage and procedure manuals. Include illustrated instructions for the individuals who will be using them.

Implement: The devices or applications will dictate their own requirements. Additional features, accessories, and auxiliary devices will need to be part of your proposal. For example, if you purchase cameras for live streaming, you will need tripods or stands to position them, as well as software for mixing, editing, producing, and broadcasting. And you certainly do not want a tangle of cables strewn everywhere! It is neither aesthetically pleasing nor safe.

Carefully plan around each initiative. Work with what you have, but for each category of devices, create a map considering the needs you have and the opportunities for improvement. **Figure 20** provides an example of such a map.

Lead: Have you ever bought a book at the library only to realize when you get home that you already have three unread copies of the same book? The best way to avoid similar situations is to maintain an accessible inventory. Create an inventory organized by categories and review it regularly. Also, pay attention to the lifecycle of devices and applications. Devices can become damaged, and software licenses expire. Your budget should include provisions for maintenance, replacement of devices, and renewal of technical support.

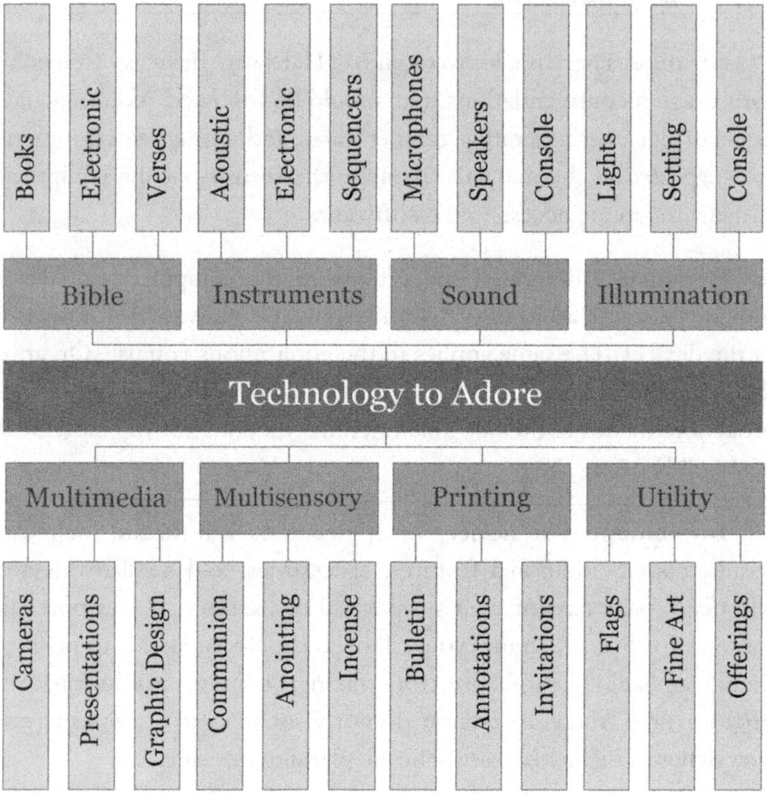

Figure 20: Technology to Adore Categories

In conclusion, technology for worship:

- Complements biblical worship.
- Puts a smile on the face of our Heavenly Father.
- Provides an augmented, media-rich, and multisensory experience.

- Systematizes elements of liturgy to provide a consistent congregational experience.
- Takes worship to new levels of beauty, exuberance, and representation of God's glory.
- Instructs individuals, families, and congregations about the importance of worship as a strategic force in mission.

And with that, we come to the end of this chapter. Do not worship technology, but worship with technology. Remember the story of the bronze serpent from **Chapter 2**. Now, let us move on to studying technology for learning.

TECHNOLOGY TO ASSIMILATE

These commandments that I give you today are to be on your hearts. Impress them on your children. Talk about them.

– GOD, DEUTERONOMY 6:6-7.

Education is the deliberate, systematic, and sustained effort to transmit, evoke, or acquire knowledge, attitudes, values, skills, or sensitivities, as well as any outcome of that effort. [65]

– LAWRENCE CREMIN, EDUCATIONAL HISTORIAN.

Several years ago, I visited an exhibition of artifacts from the colonization of the United States at the Library of Congress in Washington, D.C. There, I encountered the hornbook and for the first time understood the strange correlation between learning and pain.

When I was a child and got into some kind of trouble or mischief, my parents would say, "I'm going to read you the hornbook." That was not good news. The popular saying implied that I was about to learn a lesson through the pain of correction.

The original hornbook was a wooden tablet with a handle that had the alphabet on one side and the Lord's Prayer on the other. If you did not learn to read or pray, you learned what the wooden tablet was used for. It was better than a flip-flop slipper (dearly known in Spanish as *chancla*) because the latter did not teach you how to read or pray. I can read your mind, I know you want to rescue this ancient educational technology, but I remind you that there are now laws against child abuse.

Jokes aside, technology and education have always been intertwined since the appearance of the first cave paintings explaining methods for hunting, community life, or the relationship with divinity.

Nothing has had a greater impact on the history of education than the preservation of knowledge through language, the greatest of technologies, first through oral traditions, which were often accompanied by musical melodies or poetic rhythms as mnemonic devices, and later through writing, attributed to the Sumerians.

Writing itself has a development linked to the techniques of agrarian societies. Whether in the pictographic form like Egyptian hieroglyphics or in the phonetic alphabetic form of many modern languages, writing has served to preserve with increasing precision the economic activities of peoples in the material world with its diverse technologies.

Oral and written language served the present generation in each era and ensured the transmission of knowledge to the next, in a process that tended to accumulate increasingly more areas of knowledge and required greater complexity in language. In this regard, the European Communities, in their article *Language and Technology: From the Tower of Babel to the Global Village*, write:

> What is undeniable is that the creation of tools (technology) relies on language; indeed, it is hard to conceive of any tool, from a chisel to a CT scanner, that can be manufactured without communication, without language. What is less evident is that the development and evolution of language (its effectiveness in communicating more quickly, with

> more people, and more clearly) increasingly depend on complex tools. [66]

Hence, there is a mutual relationship: language preserves technology and aids in its subsequent development, and technology is a source of language through the introduction of new words and concepts.

The Bible relates technology and language in the story of the Tower of Babel, found in chapter 11 of the book of Genesis. The people settled in a valley in what is now the modern nation of Iraq. There, they proposed to build a city with a skyscraper, agreeing among themselves:

> "Come, let's make bricks and bake them thoroughly." They used brick instead of stone, and tar for mortar. Then they said, "Come, let us build ourselves a city, with a tower that reaches to the heavens, so that we may make a name for ourselves; otherwise we will be scattered over the face of the whole earth." (Genesis 11:3-4).

They spoke to one another and reached an agreement (language) to build with bricks and tar (technology). Notice the contrast in the passage, as it mentions not using stones and mortar, which could be easily gathered from the environment, but instead employing more advanced forms of technology that involved the manufacturing of materials prior to construction.

Technology and education have always been intertwined.

When God saw what they were doing, the passage continues, He confused their language and scattered them across the earth. Through God's action, new technologies introduced more languages and increased complexity in language and communication.

This continues to be true today. The digital age has been made possible by the emergence of specialized languages that allow us to control the behavior of computers and other electronic devices, from Machine Language and Assembly to high-level languages like Fortran, COBOL, and Java, to markup languages like XML or HTML, and many others. We can even program smart contracts for digital assets and cryptocurrencies using Solidity. Each step of digital progress has led us to create new programming languages.

The relationship between language and technology has been and continues to be crucial for educational processes. This is why our educational programs dedicate a good chunk of time to our use of language as a window to culture, science, and technology.

Educational Technology

Given the prominent role of language in the learning process, it is expected that technology would develop around language in its various forms for education. This was a gradual process spanning several centuries, but with each advancement, education has been strengthened.

The church has played a fundamental role in this regard. The early church took on the task of writing down the teachings of Jesus in the Gospels for instructional purposes. The apostolic writings were shared from congregation to congregation. Other writings not included in the Bible were also used to train early Christians, such as the *Didaché* or Teaching of the Twelve Apostles [67] and martyrdom narratives.

During the Middle Ages, the monastic church made significant advancements in the techniques of preserving and transmitting sacred scriptures. Writing became more common with the emergence of the *scriptorium*. Although initially referring to the metal stylus used to engrave on wax tablets, over time it began to be used to describe the room dedicated to reading and writing. They later adopted other materials such as papyrus, quills, and colored inks. Some monasteries specialized in book reproduction and had desks with multiple workstations, resembling our classrooms today. [68]

Another popular invention in the medieval church was stained glass windows. Since reading and writing were skills reserved for the elites who could afford the production of manually crafted books, much of the population was illiterate. Stained glass windows provided priests with a visual means to tell biblical stories, particularly those related to the birth, life, passion, death, and resurrection of Jesus Christ. Many churches incorporated stained glass windows depicting various scenes to serve as educational stations during processions.

Medieval churches also established the first cathedral schools, where theology and philosophy were at the center of all knowledge. Some of these cathedral schools eventually evolved into universities.

The invention of movable type printing press in Germany in 1440 revolutionized the medieval system and soon became a key element in education. Its immediate impact was the easy production and dissemination of books in the vernacular languages spoken by the people. The book became the king of education, with theology as its queen.

The first book to be printed was the Bible, and for the first time, scriptures could reach the hands of all. The Bible would become the banner of the Protestant Reformation and its educational initiatives. The traditional church eventually responded with its own publications of the Bible and other commentaries and study materials.

Other technologies began to emerge to support educational processes. In broad terms, in the 17th century, hornbooks were introduced, as mentioned at the beginning of this chapter, and in the 18th century, the typewriter was invented. In 1870, the magic lantern, a primitive slide projector that reproduced images painted on glass slides, was invented. The blackboard and chalk were invented in 1890, followed by the pencil in 1900, the notebook in 1920, and the ballpoint pen in 1940. The first photocopying machines were introduced in 1910, allowing the reproduction of printed materials without the need for movable type printing press.

The invention of the telegraph in 1843 and the telephone in 1877 gave rise to telecommunications. Radio appeared in 1909, followed by television and public broadcasting in 1926. The first electronic computer (ENIAC) was built by the US Army in 1945, marking the beginning of the computer age. In the following years, computers became increasingly faster, portable, and capable. The combination of telecommunications and computing through the modem (invented in 1952) ushered in the digital information age, facilitating the convergence of multiple media on personal computers until the advent of smartphones in 1992.

We now live in a world of ubiquitous digital technology that has emerged in less than thirty years. It rests upon the infrastructure of cloud computing, offers a myriad of software-as-a-service platforms for various purposes, and emphasizes task automation with robotics and artificial intelligence. This convergence of technologies is bringing us closer to what will undoubtedly be a new era in human history, the era of supersonic travel and space exploration.

The amount of technology created or adapted to support education in less than a hundred years is overwhelming. Along with this technological explosion, new methods and models emerged.

During World War II, the military, in its efforts to train new recruits quickly and efficiently, laid the foundations for educational technology and instructional technology as disciplines. Instructional technology focused on the use of technology for teaching and learning, while educational technology encompassed instructional technology as well as issues of theory, models, practices, designs, and educational procedures.

Educational technology and instructional technology are among the most developed areas in the modern era and offer great opportunities for the ministry of the church today. I am convinced that a proper combination of both has applications in the holistic formation of disciples in every corner of the earth. Although there are centers of development in these areas in Christian universities and seminaries, they have not yet been fully utilized in the ministry of the local church. There is still much room for growth.

Biblical Instruction

The Bible is an instructional manual and a learning guide. Its pages are filled with advice on how to love God and our neighbors and live a better life, with the imperative to pass on these teachings to our children and grandchildren. In this succession, Scripture serves to establish the community of faith through teaching and learning across generations and transcends all cultures.

In the previous chapter, we saw how God commanded the people of Israel to build a mobile sanctuary for worship. The desert tent became the first official educational institution because, once in service, its purpose, practices, and routines had to be communicated to subsequent generations of priests. We read:

> Obey these instructions as a lasting ordinance for you and your descendants (Exodus 12:24).

In the Bible itself, we find various ways in which these instructions are transmitted from generation to generation through rituals and traditions. Many of its narratives can be read as children's stories filled with theological significance. Other writings come in the form of repeatable proverbs that easily fit into the culture of the people. Many of its practices provide occasions and favorable conditions for learning its life principles. Even the desert tabernacle, and subsequently the temple, served as didactic models to instruct the people in matters of faith and the character of God.

The Bible is an instructional manual.

The celebration of festivals was an occasion to instruct the people in their history and identity. These were large-scale national and family events that brought unity to the nation. The most important of these festivals, the annual Passover, which commemorates the liberation from Egyptian slavery and God's saving work, came with a manual. Each festival had its own set of instructions. We see this in various passages of Scripture like Exodus 12:43: "These are the regulations for the Passover meal." To this day, Jews celebrate the Passover to instruct their children through interactive learning activities.

As mentioned before, Leviticus contains the ceremonial laws for the functioning of the temple. For this reason, many Christians have understood this book as "the law" However, the Jewish people consider it more as "the instruction." The entire Pentateuch (the first five books of the Bible) is an educational manual for the life and work of the priests and the people. Instruction itself is an instrument of God's grace, as He instructs His people in love (Psalm 25:12; 32:8; 71:17; 94:12). Thus, the prophet writes:

> Many nations will come and say, "Come, let us go up to the mountain of the Lord, to the house of the God of Jacob. He will teach us his ways, so that we may walk in his paths." The law will go out from Zion, the word of the Lord from Jerusalem (Micah 4:2, echoing Isaiah 2:3).

The book of Deuteronomy makes this instructional purpose of the Bible clear, particularly in chapter 6:

> These are the commands, decrees and laws the Lord your God directed me to teach you to observe in the land that you are crossing the Jordan to possess, so that you, your children and their children after them may fear the Lord your God as long as you live by keeping all his decrees and commands that I give you, and so that you may enjoy long life. Hear, Israel, and be careful to obey so that it may go well with you and that you may increase greatly in a land flowing with milk and honey, just as the Lord, the God of your ancestors, promised you (Deuteronomy 6:1-3).

The teaching with practical purposes is emphasized clearly, and its application becomes even more explicit in the following verses, which form the foundation of Jewish religious education and the primary commandment for every Christian. In fact, Jews learn and recite it from memory during all their commemorative occasions:

> Hear, O Israel: The Lord our God, the Lord is one. Love the Lord your God with all your heart and with all your soul and with all your strength. These commandments that I give you today are to be on your hearts. Impress them on your children. Talk about them when you sit at home and when you walk along the road, when you lie down and when you get up (Deuteronomy 6:4-7).

This commandment aligns with Proverbs 22:6, which reads: "Start children off on the way they should go, and even when they are old they will not turn from it." Now, the two following verses open space for the use of instructional technologies and educational methods:

> Tie them as symbols on your hands and bind them on your foreheads. Write them on the doorframes of your houses and on your gates (Deuteronomy 6:8-9).

This has been practiced literally in the form of phylacteries. Phylacteries are small boxes that contain portions of Scriptures, particularly this passage, and are worn on the forehead and arm. They serve as reminders that the word of God should connect our thinking with our actions (Matthew 23:3), but never as a superstitious instrument to connect with God.

An interesting fact is that the word "phylactery" itself comes from Greek, which means "amulet." The Greeks saw them as amulets, but for both Jews and Christians, amulets are considered idolatry. The secret lies not in the technology but in the Word stored in these devices to keep it in our minds and put it into practice.

During the time of the biblical kings, every socio-political and religious reform originated from the rediscovery of the Scriptures, public reading, and the practical application of its precepts. This was the case with David, Asa, Jehoshaphat, Jotham, Hezekiah, Josiah, and others, along with Ezra and Nehemiah during the return from the Babylonian exile. Note the importance of education during these periods:

> In the third year of his reign he sent his officials Ben-Hail, Obadiah, Zechariah, Nethanel and Micaiah to teach in the towns of Judah. With them were certain Levites—Shemaiah, Nethaniah, Zebadiah, Asahel, Shemiramoth, Jehonathan, Adonijah, Tobijah and Tob-Adonijah—and the priests Elishama and Jehoram. They taught throughout Judah, taking with them the Book of the Law of the Lord; they went around to all the towns of Judah and taught the people (2 Chronicles 17:7-9).

They carried the book with them. Not as we know it today, but in a primitive form of scrolls written on animal skins stored in a cylindrical case. Any initiative to bring about personal transformation and transformation within our faith communities must begin with the study and practice of the Scriptures. Fortunately, we now carry the entire book with us on our smartphones, in various versions and languages, with audiovisual study materials and a library of resources. We only lack the intention to instruct using these new tools.

Jesus' model of education was oral and peripatetic. That is, He instructed while walking with His disciples and taught them the principles of the kingdom of God from the Scriptures. His teachings were later captured by the evangelists in the form of sermons, but especially in short instructive stories known as parables. In fact, parables make up one-third of the recorded teachings of Jesus. This method, although known, contrasted with the conventional methods of His time:

> The contrast between Jesus' teaching and that of the scribes was astonishing to his hearers, as Mark reported. The escribes taught in scholarly exposition on the Scriptures, citing the opinions of earlier commentators and focusing upon subtle shadings of meaning. Jesus frequently referred to the Scriptures in his teaching and conversation, but his teaching was not about religion conventionally understood. He talked about the common experience of people as they watched. [69]

He taught in a way that used audiovisual examples from the surrounding environment to relate the truths of the kingdom to everyday life.

The subsequent generations of Christians used the teachings of the Gospels along with other apostolic writings to instruct disciples in the Christian faith, drawing from both Hebrew educational practices and integrating Greek models of education.

Assimilating with Technology

With this preamble that highlights the development of educational technologies and the emphasis that the Bible places on education and learning, I would like to share five areas for you to develop educational programs in ministry by implementing learning technologies. In other words, here are my tips for "making disciples."

Imagine taking an X-Ray of a disciple. What would you see? The body of each disciple should be made up of five bones. Yes, you guessed it: the five essential functions of mission.

Robert Pazmiño proposes an integrated model of Christian education that takes the mission of the church as the curriculum's foundation.[70] He expresses this model in **Figure 21** below:

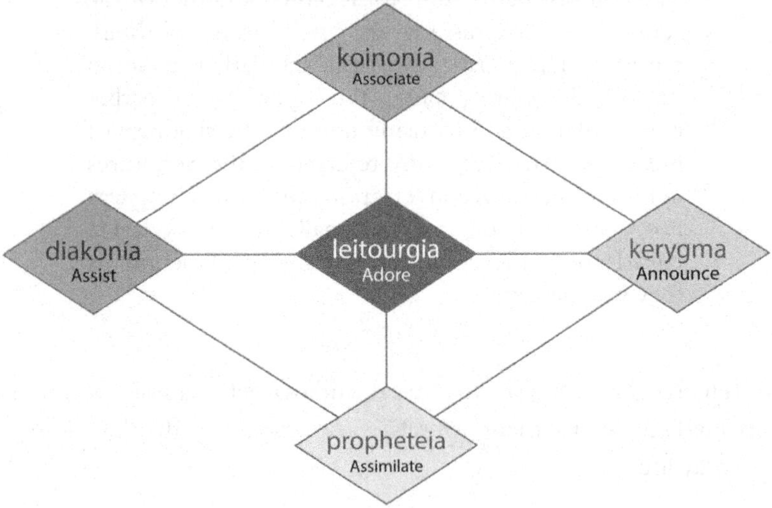

Figure 21: Church's Educational Functions

Your ministry learning programs should be organized in a way that helps each believer grow in each of the five areas of mission. Technology can assist you in forming well-rounded and balanced disciples in each area:

1. *Learn to adore*: Everything related to loving God and practicing spiritual disciplines.
2. *Learn to assimilate*: Everything related to spiritual formation and how we learn, as well as learning to teach.
3. *Learn to announce*: Everything related to proclaiming the message of the kingdom and personal evangelism.
4. *Learn to assist*: Everything related to mercy and serving others with love.
5. *Learn to associate*: Everything related to life in communion with others and the administration of the congregation.

TAIL Winds

I collaborate with an organization dedicated to training church planters. They had established an in-person program for training those who felt called to start new missions or revitalize their churches and were interested in taking the program online. We met at a coffee shop to outline how such a program would look and brainstorm the initial steps. Although there was nothing similar, we had a model to follow: we could apply TAIL to each of the five areas of technology for learning, and that is what we did.

Assimilation People

I remember telling the directors of the church planting program the phrase I often repeat: Every technology initiative should begin with people. So, we started there.

Think: Who are your Ben Jayiles and Nathanaels? Who are your teachers? Determine who your instructors, educators, and facilitators are. In the case of producing digital material, who are your producers, cameramen, and editors? Who are the graphic artists? Who will oversee the chosen technology platform? Who will promote the program through digital media? One person should oversee your spiritual formation program itself. This person should have an assistant. With them, determine the resources you have, how you will extend the invitation to participate, and which expert consultants you can hire.

Also, try to get to know the members of your group or ministry, including their level of education, study habits, immediate needs, and commitment. You can use online and offline tools, surveys, and social media for this purpose.

Did you know that instructional design specialists exist? These are people who have dedicated their lives to studying the human psyche, discovering how we learn, and applying learning techniques to teaching programs. Perhaps your program can benefit from someone like that.

Have brainstorming sessions with your faculty, creatives, and administrators to refine your objectives. Keep the contact information of all involved individuals in a shared access contact system.

Arrange: Define roles and responsibilities. Also, define the curriculum of study depending on what you want to teach. Create a recording schedule. Organize the production of each class, considering the set design and lighting. Organize the recorded material to be easily accessed and used by graphic artists, editors, and technologists. Define your delivery model.

The church planting program defined twelve sessions to be recorded on video, each delivered by an expert voice. They assigned a producer and a cameraman to handle the recordings, and an editor who divided the recordings into 4 segments each. A graphic artist created the introduction for each class. The content manager wrote the description for each course. Once all the material was ready, they hired a consulting company to create the online platform and upload all the material into a learning management system.

Have an online calendar to coordinate invitations for the recordings. Have a way to follow up with the contacts you make with each person where you can take notes. Choose a network storage system to keep all the files generated in the production process.

Technology can assist you in forming well-rounded and balanced disciples.

Implement: Plan the production activities of the study material: videos, audios, and transcriptions – do not forget about subtitles, student manuals, and teacher or facilitator guides. Plan the curation of other digital support materials. Have a clear plan for the technology you will use to distribute the education. Plan the model for online presence and reach.

We planned to launch the church planting program during an annual conference, complementing the existing in-person program that had been running for three years with the new online program for the new cohort of planters. This allowed the online initiative to leverage the momentum of the work that had been established before.

Coordinate all your activities in an online calendar and use an application to help you keep a task list.

Lead: You may notice that there are several areas of tasks to manage in the process: Curriculum (design and production), Distribution (promotion and delivery), Administration (coordination and study control), and Technology (which includes all the above). Do not forget that you will need administrators for your educational initiative.

Once the church planting program was up and running, we had to enroll students and grant them access to the learning platform, obtain progress reports from students, manage software updates, and handle video conferencing sessions, among other activities. The program included two private sessions with mentors, which also required coordination and follow-up.

Use a learning management system with functions to manage people, curriculum, educational delivery, and student progress. If you do not have a learning management system yet, I recommend that you consider it. If you are just starting out, a simple spreadsheet can store this information by function.

Assimilation Systems

We often hear the phrase "the education system," and that is because every educational initiative must first be systematic to achieve its objectives and second, it must be a system of integrated parts.

In the Bible, as I mentioned before, we see that the desert tabernacle, and later the temple, functioned as a center for instruction and spiritual formation of the faith community through ritual education. That is why it is not surprising to find a teenage Jesus learning at the temple with the teachers of the law (Luke 2:41-51). The new priests had to learn the practices of the temple to minister to God and the new generation of worshippers. Ritual education leverages technology as a system.

There are also references during the time of the kings of Israel and Judah to groups of prophets. It is likely that Samuel, during his national ministry (1 Samuel 10:10-13; 7:15-17; 19:19-24), formed these groups that later flourished during the times of Elijah and Elisha. The Bible refers to them as "sons of the prophets" (2 Kings 2:3, 15; 4:1; 6:1-4; 9:1-3), and the teachers were often called "fathers." This has led to the inference of the existence of a system of schools for prophets established in the main cities. [71]

Biblical learning systems revolve around the transmission and practice of the Scriptures, as we have seen, starting from childhood and accompanying individuals into adulthood. This became common in the modern church since the British journalist Robert Raikes introduced Sunday School in 1780. However, we must not overlook the fact that both the Jewish people and the early Christians placed a primary emphasis on the education of children and youth. The apostle Paul himself confirmed this when he wrote to Timothy:

> But as for you, continue in what you have learned and have become convinced of, because you know those from whom you learned it, and how from infancy you have known the Holy Scriptures, which are able to make you wise for salvation through faith in Christ Jesus. All Scripture is God-breathed and is useful for teaching, rebuking, correcting and training in righteousness, so that the servant of God may be thoroughly equipped for every good work (2 Timothy 3:14-17).

Let us review the learning systems using the TAIL model.

Ritual education leverages technology as a system.

Think: What are all the components of your discipleship system, Christian education, or spiritual formation? Consider the systems involved in developing the curriculum, producing educational materials, delivering classes by teachers and facilitators, the software you will use, the academic calendar, and, if it is a formal program with enrollment, the tuition collection system.

Create a written document for each learning system. It is helpful to document how each of these systems interacts with one another.

To create your basic discipleship program, you can preach a series on the five essential functions and use it as a foundation to repurpose the sermon content into various formats and channels, such as 3-minute videos for social media, 10-minute videos for a learning system, audio for a podcast, and text for study guides.

You can also incorporate content created by other institutions, such as Massive Open Online Courses (MOOCs) or materials from platforms like *RightNow Media* or *Church University*, which are constantly expanding.

Arrange: Document the components, rhythm, and interdependencies of the systems. For example, for Sunday school: meeting location, enrollment, class schedule, teachers, educational activities; in a cycle: weekly delivery with assignments and activities; depending on physical access, preparation, guides, and manuals.

In software, there is something called a disaster recovery plan. Have a contingency plan in case any part of the system fails. Organize the system components to have redundancy in assistants, teachers, materials, and even multiple individuals responsible for opening the physical or digital space.

Implement: Ensure the availability of the people involved, considering their roles and skills, the budget for spiritual formation, and the needs for future growth. Consider the short, medium, and long-term budget impact to create a self-sustainable process. Leave room to implement future projects, such as new courses, programs, or advanced levels. There are software tools available for project planning and budgeting that you can use.

Lead: Monitor the systems, their interdependencies, and their pace. Ensure that everything necessary is provided, with the right people, appropriate systems, timely delivery, and all available resources. The educational calendar is crucial, as well as having a centralized location to store documents in the cloud for each system and supporting educational materials.

Implement a change management system, which means that if there are changes in any of the systems, adjust the documentation and coordinated practices accordingly. Also, ensure clear communication of changes and adjustments through a common communication channel, such as email, phone calls, text messaging, or social media groups.

Assimilation Processes

Recently, my wife took a ten-week online class on the fundamentals of faith. The Sunday morning class met asynchronously through a social media group. Facilitators shared audio messages of the classes and responded to questions and comments from participants. Participants were required to read supporting materials, come prepared, and complete assignments for each class. The class progress, assigned materials, assignment submission, and grades were managed through a virtual classroom.

Although all the systems were free, the model never compromised the quality of education or the level of expectations for students. The class concluded with an exhaustive debate among students organized into two groups in healthy competition, resulting in a tied final score.

In the previous example, you can see the educational process in action. Despite being an online class, it fostered elements of participation, commitment, discipline, and even fellowship. The individuals connected to the class through this medium maintained personal contact and grew together in their faith.

Discipleship is not just the transmission of information. It is a process of faith accompaniment that models the values of the kingdom of God, aiming to develop Christ-like character in believers. Your educational process should lead to growth and transformation, using systems of spiritual formation to influence the lifestyle and character development of your disciples.

Instructional Design, by the way, views education as a systemic process with planned activities, material resources, and conducive environments that interrelate to facilitate the transmission of knowledge. It would not hurt to incorporate the concepts of this discipline into our discipleship training programs, but that would be material for another book.

Discipleship is not just the transmission of information.

Think: Document the learning processes with flowcharts. Some key processes for learning include educational material production, maintenance of content on the online platform, student progress tracking, class delivery, and the student lifecycle.

Ensure to note the difference between training (to learn how to perform repetitive tasks), instruction (to shape character and values to respond to daily life circumstances), and nurturing (to provide guidance and modeling of behavior).

Arrange: Arrange the workflow by defining tasks, actors, and decisions to be made at each step of the process. This is a good opportunity to integrate with other processes, automate repetitive tasks, and optimize systems. Document the workflow using appropriate applications.

If your education model includes the development of audiovisual content, make sure to model the entire production process, from content evaluation to teaching and reviewing its impact. I like to use what I call the content machine for this purpose, as illustrated in the following image.

Figure 22: The Content Machine

When it comes to delivery, explore multiple channels for teaching and remember that nothing is out of reach. You can implement a Learning Management System (LMS) or create a subscription website with Content Management System (CMS); both have free and open-source systems that can benefit your in-person and online programs.

Implement: You will always have the perfect flow if everything goes well without any possible deviations. But what if it does not? Plan workflow contingencies to consider those exceptions.

Plan and monitor the workflow. Keep the process simple. Do not make these activities so complicated that you cannot follow them. Remember, less is more.

Lead: Look for opportunities to improve learning processes. Identify inefficiencies and apply innovation to enhance existing processes.

Handle everything well. Together with your team, determine which learning processes are most critical, such as Sunday School, ministerial training center, or discipleship, for example. Create the procedure manual and keep the documentation up to date with task diagrams.

Assimilation Policies

Think for a moment: education depends so much on policy that there is a department or ministry of education in national government cabinets. This should tell us something about education and technology policies for learning in the ministry. More formal training institutions must adhere to many of these government policies.

Education policies in the ministry impact various areas of the education process, from teachers to students, and from content to education administration. The formality or informality of education in the church, Bible institute, university, or seminary largely depends on political considerations.

The following brief list can give you an idea of the type of things to consider when establishing a spiritual and ministerial training program:

- The requirements to be a teacher or facilitator.
- Continuing education requirements for educators.
- Models of educational administration.
- Curriculum or syllabus.
- Admission requirements for new students.
- Minimum effort for student retention.
- Privacy of student data.
- Teaching and learning practices.

- Preservation of academic records.
- Government laws regarding education.
- Eligibility for official accreditation and certifications.

In the United States, there are several government agencies and theological accrediting associations, each with their own requirements for membership. It would not hurt for your local church programs, ministerial institutes, or online initiatives to consider affiliating with these agencies and organizations. [72]

In the same passage we previously reviewed for deacons, the Apostle Paul gives his disciple Timothy a list of the qualities of bishops that could very well be applicable to teachers and administrators of Christian education.

> Now the overseer is to be above reproach, faithful to his wife, temperate, self-controlled, respectable, hospitable, able to teach, not given to drunkenness, not violent but gentle, not quarrelsome, not a lover of money. He must manage his own family well and see that his children obey him, and he must do so in a manner worthy of full respect (1 Timothy 3:2-4).

This passage is a good starting point for establishing your educational policies.

Think: What are the policies that govern teachers and learning technologists in your church? What are the best practices you can emulate in educational matters? Every decision you make within the framework of learning technology has political implications that you must consider.

Write the policies for the use of technology for learning and share them in visible places for everyone. Make them official practices. Be flexible enough to adjust these policies according to changes that arise with use or new legislation.

Arrange: Familiarize yourself with local laws and accrediting agency requirements. Include legal experts if necessary, depending on the scope of your spiritual and ministerial education program. Consider legal protection and copyright protection for the content you create for your programs, especially if they will be exposed to the general public.

Organize the policies by categories. Educate participants about these policies. Publishing some policies on your website may be mandatory in your context. Do not overlook this.

Implement: Plan review sessions for your policies. Plan the process for accrediting your educational programs. These processes take time and, depending on the level you want to achieve, require capable people to guide you. Anticipate the impact that technological changes may have on your policies and vice versa.

Use a calendar to communicate policy and procedure review dates. Establish educational affiliations and stay informed. Participate in conferences and conventions on educational policy so that you can make relevant and timely adjustments.

Lead: Stay prepared for change. Manage and control policy and practice adjustments. Ensure they are followed; be a guardian of your policies.

It is not about becoming a legal expert, but follow the laws of the land, such as the Elementary and Secondary Education Act, the Family Educational Rights and Privacy Act (FERPA), the Higher Education Act, and the Individuals with Disabilities Education and Rehabilitation Acts.

Also, guard the foundations of the gospel to preach appropriate sound doctrine with integrity (Titus 2:1).

Assimilation Devices

As we have seen, Christian education and formation revolve around reading, studying, and practicing the Scriptures. The Bible is central to spiritual instruction, and study resources and teaching methods are created around it.

In addition to this, learning activities take place in the church and in dedicated ministry education venues through ritual education and various forms of discipleship. The advent of digital technologies opened possibilities for new models of multimedia, online, distance, and hybrid education.

For all these models or combinations, there are hardware and software tools that you can use in the construction of knowledge and the transmission of Christian character. The possibilities are vast. Let us review some ideas.

Think: What educational tools are you currently using? How are you addressing the educational needs of your church or ministry? Are they aligned with your strategic plan?

Gather educators and administrators to think about the devices and software needed to complement your educational and training initiatives. Do not forget the resources you need to cater to each generation in their stage of life, meaning, do not neglect the instruction of children and youth. Keep in mind that they are more familiar with digital technologies; some disciples are digital immigrants, while others are digital natives.

Arrange: Organize education in the church with worship and preaching practices, and with classroom instruction. Each environment has its own needs for educational devices. Organize the tools necessary for the administration of your educational programs. Organize the security of equipment and secure access to software. Productivity software can greatly assist you in the organizing process.

Also, organize all the devices you will need for the implementation of your educational program. When I was leading the discipleship program at the church, we implemented flat-screen smart TVs in the classrooms. Then, we installed wireless screen mirroring devices with HDMI cables to the TVs, and these devices were connected to the church's wireless network. With everything set up, we trained teachers and facilitators to share class presentations from their smartphones or tablets without the need for computers. More details in **Chapter 11**.

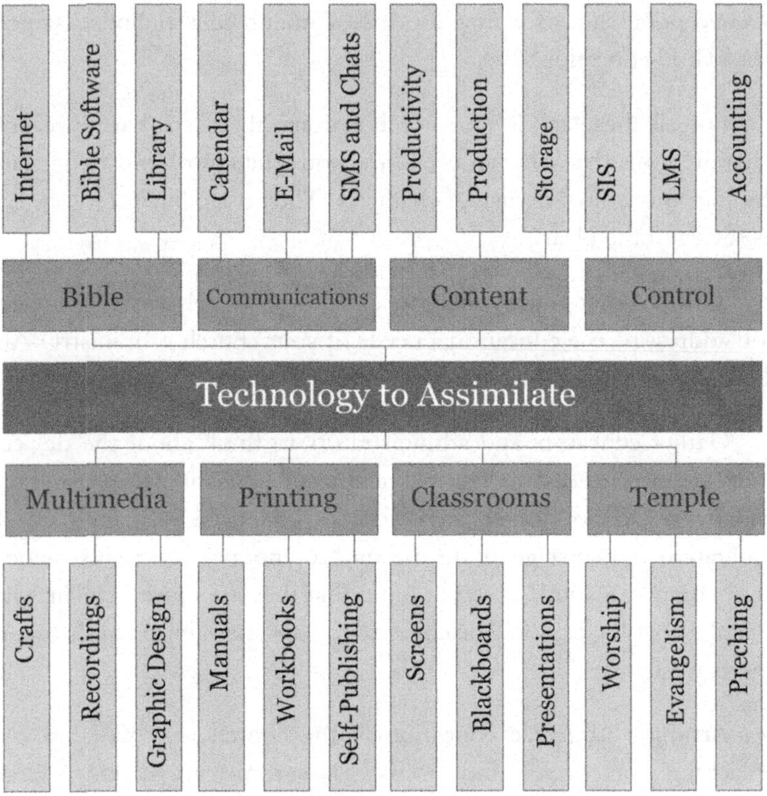

Figure 23: Technology to Assimilate Categories

Implement: The needs and objectives of your program will dictate the use of auxiliary educational technologies. Even if you do not have all the pieces in place, plan with a future perspective.

Plan for each category of educational devices considering the opportunities to form character-disciples who are knowledgeable of the Scriptures. **Figure 23** is not exhaustive, but you can use it as a guide to organize the learning devices you will employ.

Lead: Control and coordinate each area to ensure smooth operation. Practice regularly to fine-tune the details. It may seem overwhelming, but remember, one step at a time will take you far. Therefore, prioritize each area and plan in stages.

You have the option to manage your programs with specialized software for Learning Management Systems (LMS), which can be adapted to handle student progress, or you can implement a specialized Student Information System (SIS) to manage student information. Refer to **Chapter 11** for more information on this.

For twenty years, I have been a specialist in this kind of software, which requires advanced skills in managing commercial databases, application servers, web servers, and a combination of programming languages and technical platforms. I have been able to provide consulting services to various universities, seminaries, and educational institutions in the selection and management of these systems. Some accreditation programs require the use of such systems before granting their seal of approval.

Technology for learning benefits you in significant ways:

- It provides tools for spiritual growth.
- It coordinates resources to address educational needs.
- It organizes and distributes educational material.
- It produces healthy and knowledgeable disciples in the faith.
- It trains saints for the work of ministry.
- It works for you even when you are not teaching.
- It brings education to places you may never personally visit.

We do not have to "read the hornbook" to anyone, but it is our duty to expose members of our ministries to educational programs that make use of these and other technological advancements. My prayer is that new models of disciple formation will emerge from this newfound understanding to cultivate well-rounded disciples.

In our breakdown of ministerial technologies, it is now time to address technology for outreach.

TECHNOLOGY TO ANNOUNCE

Go into all the world and preach the gospel to all creation.

– JESUS CHRIST, MARK 16:15.

The highest form of worship is the sound of consecrated feet going to reach the lost and the needy.

– BILLY GRAHAM, AMERICAN EVANGELIST.

We descended the stairs of Hiram Bithorn Stadium in San Juan, Puerto Rico, heading towards the field near the pitcher's mound. In front of us, on a stage set up in center field just behind second base, a choir of over two thousand five hundred voices sang hymns of worship to the accompaniment of a full orchestra.

The atmosphere was vibrant, filled with prayer and great anticipation on that warm March night. We were preparing for a grand celebration. We were directed to the seats on the green field as over twenty-three thousand people filled every corner of the stadium.

It was Billy Graham's final major crusade in Puerto Rico. The year was 1995. Graham had visited the island in 1958 and again in 1967, but this time was very different. The technological display was ubiquitous and unprecedented in the history of world evangelism.

Wherever you looked, there was technology at a level surpassing the Olympic broadcasts. One-hundred sixty Beta Cam D2 cameras were used to cover the campaign from every angle. They were joined by modern cherubs hovering in the air with their propellers. The live broadcast would be delivered via closed-circuit television by 30 satellites to over 175 countries around the world, to several local television networks, and on several 40-foot screens strategically placed.

At the same time, the transmission would be received simultaneously at the adjacent Roberto Clemente Stadium through a mobile television control room with 61 monitors and 8 television cameras. This allowed the control center with 25 booths to edit the transmission in 4 languages at once and to be translated into 102 languages from the 48 translation booths, each with a translator and an alternate.

Other modern advances developed for the occasion ensured that the message of the Gospel reached over 170,000 people in person and over 8 million people worldwide. It was unheard of at that time. Nearly 2,500 churches collaborated with the effort, and among them, half a million volunteers were trained to receive the new converts, who numbered 3,149 people, mostly young individuals, in Puerto Rico alone.[73]

Billy Graham then said:

> The technological revolution has turned the world into a global village... It is time for the Church to use this technology to tell the world that amidst the chaos, emptiness, and despair, there is hope in Jesus Christ.

This massive technological effort for global evangelization caught the attention of Technology, Entertainment, and Design (TED), who invited Graham to deliver a keynote speech in February 1998. Twenty-five years have passed, and Billy Graham is no longer with us, but his words about using technology to communicate the powerful message of the gospel of Jesus Christ are still relevant today.

Technology for reaching out refers to technology dedicated to proclaiming the gospel. In the case of the crusade, we can mention the twelve television production trucks, satellite links, closed-circuit television for global transmission, and broadcasting, among many other things.

Perhaps you may not have direct access to some of these technologies, which are often expensive and require high maintenance and technical expertise. However, today you have affordable access to outreach technologies that were not available twenty years ago. With so many advanced technologies at your fingertips, what cannot you do today to reach those around you? Are you ready to reach your "global village"?

Go!

In **Chapter 5**, we clearly defined the mission of the church. At the heart of the mission are the two mandates of Jesus Christ that I mentioned earlier in Mark 16:15: Go and preach! and Matthew 28:19: Go and make disciples! In both cases, the imperative is to go. It is unfortunate that many churches I have encountered in my faith journey have inverted this active command and turned it into a passive call: "Come!"

The church that Jesus envisioned was not a church of "come" but a church of "go." This shift in focus should be reflected in everything the church does with technology when proclaiming the message of the gospel. Let me explain with an example.

In my younger years, I was asked to manage the bulletin board for the church's youth group. We are talking about the 80s when computers and portable printers were not readily available to everyone. I inherited letter stencils to fill with markers on bond paper and *Elmer's Glue-All* to adhere magazine clippings and organize announcements for upcoming activities. But few people found out because few bothered to read the bulletin board. The technology was primitive, perhaps that is why it had a very poor effect.

So, I put all my creativity to work by creating raised representations of Bible passages using mixed media techniques and unconventional materials and paints. In other words, I improved the technology. This increased traffic to the bulletin board because the novelty generated interest. But the interest was short-lived. I put in a lot of effort. I even recruited other talented young people. But despite all our efforts, the youth group did not grow.

Some ministries treat their digital media like a hallway bulletin board. They put a lot of effort into announcing their activities, sometimes with a poor appearance of overused templates, and other times with mixed media that surprises those who see it. In fact, they may be using new digital bulletin boards in the church hallways. The problem is that few people see them.

If you feel the same way, the solution is to go. We went to the streets, to the squares, to the neighborhoods, to the shopping centers, to the schools, and to the prisons, and the more we went, the more people came to the youth meetings —except for those in prison, of course.

Change your perspective. Think of your website, your social media, your email system, your graphics, and your videos, among other things, as ways for you to go and not just for people to come.

Evangelizing the Cyberculture

Another challenge present in evangelism today is the influence of technology on culture and the extent to which new generations of digital natives enter the public discourse, creating a culture of cyberspace or cyberculture. It is a new digital environment of social interaction mediated by information and communication technologies that is developing its own values, symbols, language, economy, and therefore, its new existential and epistemological metaphors.

While the technological church may have a sign like the websites of old that reads "Under Construction," the cyberculture is not sitting around for the metaversal church to be ready. The powerful message of the gospel demands that new digital ministers understand life in virtuality and interpret that message in the new worlds.

It is not just about using today's technology in the worlds of yesterday. We must develop different strategies for each digital country, many of which have billions of real and virtual citizens and hundreds of thousands of diverse groups, each with a dominant cyberculture but also many cyber subcultures.

> *The cyberculture is not sitting around for the metaversal church to be ready.*

I explore this topic in more depth in the third book of this series titled *Cybertheology*, but I am confident that some of the advice in this chapter will help us decipher what is most beneficial for us to do in our contexts regarding reaching out with technology.

Announcing with Technology

Technology for outreach has the potential to make evangelism more accessible, convenient, and attractive. It provides an opportunity for us to share the love of Christ with a wider audience and have a positive impact on people's lives. Below are some of its advantages:

- *Increased engagement*: Technology can elevate our sense of common mission, helping us to increase people's engagement in sharing the message of the Gospel. Through virtual events, webinars, and live streams, individuals can actively participate in evangelistic efforts.

- *Reach*: Technology allows us to connect with people from different parts of the world. Through the internet, we can reach individuals who may not have had the opportunity to hear the Gospel before and stay connected to our community from a distance. Additionally, a simple search can provide us with many resources that we can use to address their needs.

- *Cost-effectiveness*: Today's technology can be used to evangelize at a lower cost than traditional methods. For example, using social media platforms to share the message of the Gospel is much more economical than printing and distributing physical materials or paying for satellite television.

- *Convenience*: With technology, people can access the message of the Gospel anytime and from anywhere. Online resources such as podcasts, videos, and e-books make it easier to learn about Jesus Christ and grow in faith. Mobile applications can help you share the Gospel personally without having to be an expert, turning every person into a mobile evangelist.

- *Interactivity*: Technology provides an opportunity for interactive evangelism that resembles faith on the go. For example, online discussion forums can be created where people can ask questions and receive answers from other believers. The key is to use this interactive capability to establish genuine conversations and lasting relationships.

TAIL Winds

I have had the privilege of serving various churches in technology matters for several years, but nothing compares to the year 2020. Suddenly, I began receiving calls from pastor friends and referrals requesting help in different areas. Some wanted to start online support groups, given the surge in public awareness of virtual meetings. Others were trying to implement online payment portals to receive tithes and offerings. And many wanted to strengthen their digital presence by creating or revising their websites and social media accounts. But none of these areas were as popular as the need to livestream church services. Many were not prepared to make the cyber leap and had not even considered the possibility, let alone met the technical and tactical requirements. However, the circumstances called for an immediate response.

From personal experience, livestreaming church services brought worship and clear preaching into my living room during times of great confusion. For several months, livestreamed services were our only connection to the local church. It was an event that forced the church to return to its historical roots: the home church.

In the urgency of the situation, I devised a plan to help churches respond. But the response would come in terms of the Ministry Technology Matrix and the TAIL model. You are already familiar with these models.

Announcing People

Question number one: Who was or would be responsible for the livestream? This is the fundamental question because, due to the nature of the technology, the pastor himself could not be in front of and behind the cameras at the same time. It sounds almost comical. Pastors are known as one-man bands in some of our churches, but in this case, livestreaming without delegating tasks was very challenging.

Think: Who are your evangelists today? Who are your witnesses? Who else do you have to send? How are you going to go? Jesus spent time at the beginning of His earthly ministry identifying twelve men. He walked with them to show them what they were to do, and then He sent them out in pairs to do the same.

> Calling the Twelve to him, he began to send them out two by two and gave them authority over impure spirits… They went out and preached that people should repent (Mark 6:7, 12).

He then chose 72, even saying these were not enough:

> After this the Lord appointed seventy-two[a] others and sent them two by two ahead of him to every town and place where he was about to go. He told them, "The harvest is plentiful, but the workers are few. Ask the Lord of the harvest, therefore, to send out workers into his harvest field…" (Lucas 10:1-2).

If Jesus delegated evangelism, we should do the same. Delegation applies to every evangelistic initiative, including digital evangelism. God is calling us to take a technological leap in evangelism, and in this, we are also guided by the Holy Spirit, who was given to us without restriction:

> For the one whom God has sent speaks the words of God, for God gives the Spirit without limit (John 3:34).

Do as Jesus did. Identify the people of influence, evangelists, preachers, and community missionaries. Form your team for in-person and digital evangelism to work in a coordinated effort. Maintain a regular rhythm of work. But above all, let them go!

Define your community, its demographic composition, and the institutions it has. Walk the streets with surveys on tablets to get to know them and as an excuse to connect with people and their immediate needs. You can also use online resources to learn about your target community.

Do not neglect the communities of cyberspace on which you can exert your influence. To reach the digital world, assign someone responsible for an editorial calendar that considers multiple media and maintain a community management team to sustain interest and commitment to the message of your ministry. If you do not have someone on your team, consider hiring expert consultants to help you define strategies.

Arrange: Define the responsibilities and roles of each person. Document everything. Take advantage of the tools mentioned earlier to stay organized. Have a routine of meetings to report on your progress and challenges along the way.

For example, to livestream high-quality services, you will need individuals with skills and knowledge in at least the following areas:

1. *Video Producers*: Someone who can direct and coordinate video production, establishing shots, angles, and proper lighting.

2. *Set Designers*: One or more individuals with a creative eye for spatial design, envisioning the stage setup and preparing the broadcasting environment. Explore indoor and outdoor spaces to take the church beyond the four walls and traditional formats. Some churches have created a dedicated space to address online viewers during the service, with the video feed connected to a console that alternates with the general service feed; you could test this in your ministry.

3. *Camera Operators*: One or more individuals who can handle the cameras and ensure that the right shots are being captured following the producers' guidelines.

4. *Sound Technicians*: Someone who can monitor and adjust the sound during the livestream to ensure good audio quality. Make sure your streaming audio is of the highest quality. Most people watch streams while doing other activities like driving

or exercising. In these cases, audio is more important than video. Poor audio will bore people and make them disconnect.

5. *Video Editors*: Someone who can edit the video to produce an excellent final product. Editors may create opening and closing graphics, adding a church-branded overlay or video announcements.

6. *Livestream Coordinators*: Someone who can set up and manage the livestream on a livestreaming platform, social media platforms, or multiple channels.

7. *Social Media Managers*: Someone who can handle the church's social media accounts to promote the livestream and respond to viewer comments and questions during and after the broadcast.

It is important that these individuals have the experience and skills necessary to ensure that the livestream is of high quality and that the visual and audio experience is appealing and enjoyable for viewers.

Implement: Consider the following aspects to ensure that team members in both in-person and digital evangelism are well-trained to share the gospel:

1. *Spiritual Preparation*: It is important that those involved in evangelism have a close relationship with God and a solid knowledge of Scripture. This will enable them to share the Word of God clearly and effectively.

2. *Understanding the Target Audience*: Before engaging in any form of evangelism, it is necessary to understand the target audience. This includes their culture, language, beliefs, and needs. This way, the gospel can be presented in a relevant and meaningful way.

3. *Evangelism Strategies*: There are different ways to deliver the message of salvation to people. When planning the involvement of individuals in evangelism, consider the various strategies that can be utilized, such as personal evangelism, group evangelism, and evangelistic events, among others. Adapt these options to the digital world and make decisions accordingly.

4. *Teamwork*: Evangelism is a task that requires teamwork. It is important for those involved in evangelism to work together in harmony and unity, sharing ideas and supporting one another.

5. *Evaluation and Follow-up*: After each evangelistic activity, it is important to evaluate the results and follow up with those who have shown interest in the gospel. This will allow you to measure the impact of evangelism and continue supporting individuals in their spiritual journey.

Consider religious and festive calendars to coordinate the outreach ministry. Certain dates represent better opportunities for people to hear a message of hope and love, such as Valentine's Day or Mother's Day. These dates should be part of your annual coordination, along with meetings that address the spiritual preparation, audience, evangelism strategies, teamwork, and follow-up. Unity within the team and effective coordination will maximize the impact of your actions.

Lead: Administer the evangelism and outreach team. Use regular meetings to understand the challenges faced by those serving on the team, both personally and in the execution of their duties, and provide support and prayer. Remember that ministry is done together.

Communication is key. Decide which communication channels you will use with your team and for proclaiming the gospel, considering who will be in charge, even if it is Sister Emma making phone calls. Be clear in your communication.

Announcing Systems

Some churches had to act quickly and with very little budget. The good news was that they no longer needed 160 Beta Cam D2 cameras, 61 monitors, 30 satellites, or 25 editing booths, as in Billy Graham's campaign. Some churches took advantage of the capabilities of their smartphones to go live directly on their social media platforms.

Without giving much thought to the quality of the transmission, let alone the relevance of the content being transmitted, churches went online in an instant, reaching the global village within minutes and creating remote communities "to the ends of the earth." This agility also became a disadvantage, with poor transmissions of dark and blurry images due to lack of lighting, very poor sound with ambient noise and echoes, inappropriate scenes due to attempts to reproduce the pulpit in the living room, and monotonous topics without impact on the cyberculture. These things allowed us to open our eyes to new realities.

Consider the advantages and disadvantages of your outreach systems and complement them with each other. Review the weaknesses that you can observe and bring them to the discussion table to address them.

Strengthen your digital presence so that people can find your website and make it attractive. Fine-tune your message on social media to be consistent with your ministerial identity. Also, think about the penetration of mobile devices and how you can have a presence through them. For example, you can turn your sermons into a podcast - an audio streaming system using a syndication subscription model (RSS, Real Simple Syndication), or into a blog that publishes written content using the same RSS technology. Or perhaps creating an internet radio station would be suitable. If that is the case, be creative in the content and format of your programming.

Think: Review the critical outreach systems you have and determine which ones you need to initiate or strengthen. This becomes much easier if you distill it from your own strategic planning. You should strive to align all your systems with your strategic objectives and potential impact.

Establish which evangelism method you will use and train your team in its use. The method is a system. If necessary, create your own program tailored to the needs of your context.

Research what you need and what is available within your budget and technical capabilities. Do not worry if there is a knowledge gap because that can be filled with training, often provided by the same provider. Choose two to three options in each category and try them all to select the best one. One of those key systems could be the live streaming example used in this section, but the same principles can be applied to your other methods. Moreover, there are traditional methods that now offer digital tools (**Chapter 11** under "Evangelism"). Explore, research, review, and decide.

Strive to align all you systems with your strategic objectives.

Arrange: Review the section on systems in **Chapter 3**. Note the objects, order, and objectives of each system you have considered. Document how these systems relate to each other, with special care in noting interdependent systems.

Review how each system could fail and minimize the risk. We have already discussed the importance of maintaining a set of comprehensive and accessible documentation for others to step in when someone is absent.

Implement: Dare to try out various systems in each category before committing to one. It is okay to do so. When we started, we tested several online streaming platforms, webinar platforms, and live meeting platforms, some free and some paid. Plan the costs; each platform has its own price and its own form.

Explore your options, test, and select the systems that best fit your needs and budget, and that are easy to use for members of your ministry team.

Lead: Managing the outreach systems is crucial to ensuring their effectiveness and maximizing their utilization. Clearly establish the maintenance of the systems and control the dissemination of your content through them.

You should always define the guiding thread of your evangelism and outreach campaigns. Once defined, decide how your systems fit into and coordinate with that guiding thread. For example, the weekly worship service provides a good rhythm and serves as a guiding thread for your online streaming. Make sure to have one person in charge of each system and a second person serving as a backup in case the primary person is unavailable.

Each platform has its own price and its own form.

Evaluate your own live broadcasts with a critical eye, as there is always room for improvement. When assessing your systems, you can use the SWOT analysis. SWOT stands for Strengths, Weaknesses, Opportunities, Threats.

Both internal and external factors will impact your ministry's efforts. Note that Strengths and Weaknesses are internal aspects of your ministry that you can control to some extent, while Opportunities and Threats are external factors over which you may not have direct influence. Become a good manager of the things you can change.

Figure 24 on the following page shows the SWOT analysis we conducted for a church's live streaming program. I hope it serves as a guide for you. Once you complete your analysis, focus on your strengths and opportunities, rather than your weaknesses and threats. Strengthening your strengths will mitigate your weaknesses.

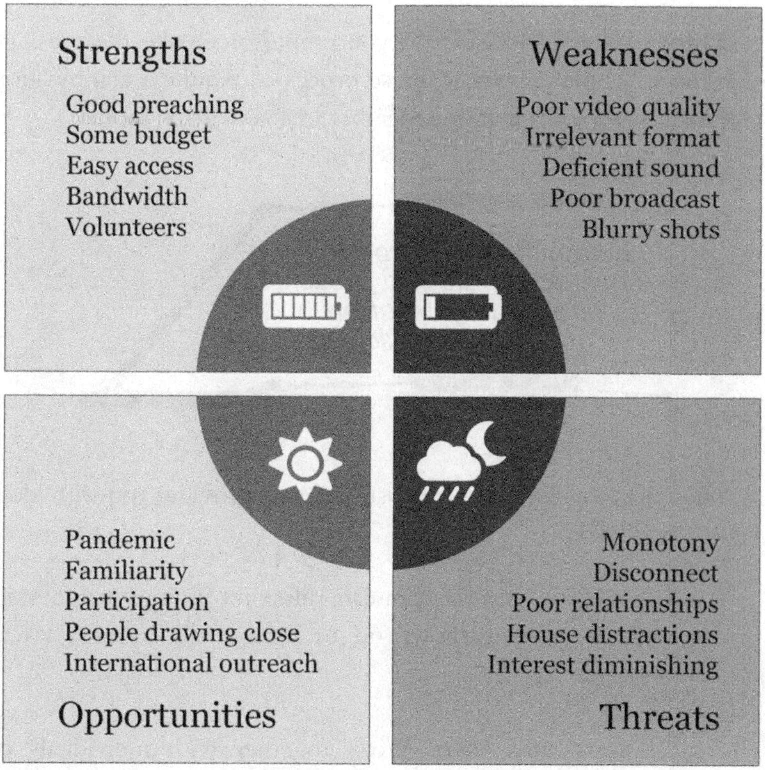

Figure 24: Live Broadcast SWOT

Announcing Processes

Remember: movement and transformation. By definition, evangelism is a process of movement — "go and preach!" and transformation — "go and make disciples!" Many programs I have encountered focus only on the announcement—the hallway bulletin board—without discipleship. Disciples are not made without transformative intention and committed mentoring.

You should ask yourself: "How do I want to mobilize the people I reach and what kind of transformation do I want to see in their lives?" Draw a map of conversion to mature discipleship. Use familiar images. For example, Rick Warren uses bases on a baseball field. [74]

Think: What processes can you employ to make the message reach more people? Typically, these processes require a combination of strategies and tools working together in a systematic model.

Evangelism is a process of movement —"go and preach!" and transformation —"go and make disciples!"

The following list is not exhaustive, but it provides you with ideas to get you started:

1. *Receive online donations*: Stimulate mission offerings with mobile applications that allow you to receive donations via text messages.

2. *Collaborate with others*: Work together with individuals or organizations that have similar goals to expand your reach. Collaborate on content creation, organize events, or partner with other churches, schools, or family therapy offices for outreach activities.

3. *Create quality content*: Share high-quality content to increase the reach of your message. Use videos, images, infographics, or articles to provide informative, useful, and relevant content to your audience. Think outside the box and offer ideas beyond church activities. For example, provide tips for quality family time or saving money during holidays by giving meaningful gifts.

4. *Utilize social media*: Social media platforms are excellent tools for spreading your message to a large audience. Use relational, specialized, or professional networking platforms to share your message and reach a wider audience.

5. *Utilize email marketing*: Collect email addresses and phone numbers to share newsletters or send personalized messages to your followers via text.

6. *Invest in advertising*: Paid advertising can be a useful tool for promoting the gospel. Online advertising platforms, search engines, and social media platforms allow you to define your target audience and display ads to people who meet specific demographic, psychographic, or geographic criteria.

7. *Connect with traditional media*: Television and radio, although now considered traditional media, are still effective for proclaiming the gospel.

Arrange: Take the list above or create your own tailored to your context and organize all the relevant areas. Coordinate the delivery of high-quality content in the chosen formats, from ideation to production and publication. Utilize your drama or dance group to create creative and high-quality videos, for example. Allocate your budget to create a paid advertising campaign on search engines. Some platforms offer grant programs for nonprofit organizations to obtain a significant amount of advertising money for free. Foster collaboration and relationships with other ministries to reach the city with a kingdom mindset.

Implement: Develop plans following the established and organized processes aligned with your mission to mobilize and transform. Mobilization is the first step in outreach, but transformation makes the follow-up process critical. For example, if your intention is to have everyone in your ministry participate in social media evangelism, plan:

1. Content creation for individuals to share on their social media and groups, providing them with the content and instructions on where to download and share it.
2. Dates when everyone will flood social media by sharing the same content.
3. Contact channels to capture engagement, such as direct messages on social media, email, or phone.
4. Follow-up strategy for engaging with people who interact with the shared content.

Lead: When your outreach processes are in motion, you will discover things along the way that need to be managed. Details may be overlooked, and system failures may need attention. Even positive problems can arise when your work is extremely successful, and your followers grow exponentially. Manage growth and grow in your administration.

I remember a time when our digital congregation grew so much that we could not fit in our virtual temple anymore. We went from less than five hundred subscribers to two thousand in just six weeks, surpassing the limit of our free email marketing software. We had to transition to a paid model to serve so many people.

Manage how you grow and grow in how you manage. This means that your people continue to grow in knowledge and capacity as your ministry grows and matures.

Announcing Policies

What causes Christians to be persecuted, tortured, and even killed for proclaiming their faith in certain parts of the world? It is the policies of those places. I know it sounds drastic, but it clearly shows that there are ministry outreach policies that we cannot overlook. In countries where this is a reality, some mission agencies broadcast the gospel from border radio stations, smuggle Bibles, sometimes even individual Bible pages, or gather in secret.

Be thankful if you live in a nation with freedom of speech and religious liberties that open a world of possibilities for sharing the gospel. However, do not let the multitude of options paralyze you from acting. Use policies to your advantage, or rather, in favor of proclaiming the gospel.

Think: Write down the policies for using technology in outreach and discuss them in your prayer, recruitment, and planning meetings. Document the requirements for joining evangelism teams, the necessary training, and participation in meetings and outreach activities. Define sound doctrine in preaching. Do not forget to write down the values for interacting with others and adhere to the guidelines of your corporate identity.

By the way, create a corporate identity manual or branding guide. Your corporate identity is an outreach policy as it will shape how people perceive your ministry. Adhere strictly to your corporate identity.

Your corporate identity is an outreach policy.

Your corporate identity manual is a document that establishes guidelines for the proper use of identity elements with the aim of maintaining a consistent and coherent image across all communication channels of your ministry. The components of a corporate identity document may vary depending on your organization, but it typically includes visual and verbal elements that represent your ministry's brand. It usually includes:

1. *Logo with approved variations*: This is the visual representation of your ministry that identifies you.

2. *Official corporate colors*: These are the colors that have been defined to represent your brand and are used consistently in all published materials.

3. *Typography*: This refers to the set of fonts and characters used in your written communication. The typography should be consistent with your brand image.

4. *Slogan*: This is a short and memorable phrase used to reinforce your message and incorporate your values.

5. *Brand motto or Bible verse*: This is a Bible verse that fully identifies your ministry according to your strategic plan.

6. *Corporate stationery*: This includes business cards, letterheads, envelopes, folders, and other elements used for official communication.

7. *Web design*: The design of your website should be consistent with your ministry's corporate identity, using the same colors, typography, logo, and tone of communication.

8. *Graphic templates*: These templates not only facilitate creative work but also provide consistency to the communicated message. You may have templates for several purposes.

9. *Advertising*: The advertising we mentioned earlier should reflect the corporate identity, using the same visual elements and messages as other company materials.

10. *Tone of communication*: A document that defines the tone of language and audiovisual style to be used, including a description of your target audience, your communicational objectives, the personality of your brand, the emotional tone, and the language style, among other things.

In this list, you can see the relationship between your identity manual as an outreach policy and a technology tool at the same time. For more information, refer to the references in **Chapter 11**.

Arrange: All your policies should be present, visible, accessible, categorized, and current to avoid any surprises. For online broadcasts, consider reimagining the content. Some churches have transitioned from simply livestreaming the Sunday service to becoming more like a rich audiovisual magazine, with live and archived worship, participation of people from around the world in prayer or preaching, and short messages relevant to the times.

Implement: Consider everything that can go wrong and prepare your policies to protect yourself. The gospel is content that communicates the Word. Therefore, outreach through technology relies on the creation of content, whether written or audiovisual, which is subject to copyright regulations. Most social media platforms have safeguards against the use of content that you do not own or do not have permission to use. If you violate these rules, they have the option to close your channel or even your account.

Do not engage in piracy or plagiarism. Do not use content from others without permission. Do not download any random photos from the internet or use music from others in your videos unless you have written licenses that grant you the right to use them. You may not be aware, but even quoting verses from your preferred Bible version in commercial material can be considered a copyright infringement. Check the websites of publishers to ensure you comply.

The gospel is content that communicates the Word.

Also, protect yourself against content usurpers. Create clear policies stating that all content generated by your managers, graphic artists, writers or copywriters, and actors is the property of the ministry or organization and not the individual creators. This will give you the right to make modifications as necessary without repercussions for the creative members of your team.

Your website should express its usage policies. If you offer products for sale or donation, you must protect the information you collect. You must also protect the personal data of those who share their email addresses and phone numbers with you. There are laws in Europe (GDPR) and California (CCPA)[75] that are highly sensitive to the handling of this information. If you operate in these areas, you must ensure explicit consent for the use of *cookies*[76] and the collection of personal data.

Lead: Work retrospectively to update your technology outreach policies and display them in relevant mediums. Then ensure that these policies are consistently followed and regularly updated.

Pay for music usage licenses. Annual licenses to use Christian music are not as expensive as you might think and depend on the size of your church or ministry, the type of usage (e.g., in worship services, rehearsals, live streaming, recorded or printed materials), and will vary with changes in your ministry. In the United States, you can acquire licenses through organizations such as Church Copyright License International (CCLI), ASCAP, or BMI.

Pay for audiovisual resource licenses. There are specialized websites for using photos, graphics, icons, and videos while respecting copyright. Refer to **Chapter 11** under "Licenses" for more information. You can also use freely distributed resources through Creative Commons if you adhere to their attribution guidelines.

If you do not want to spend money, produce your own materials, or I invite you to check out my article *7 sources of free photos to use in your visual presence*[77] for more ideas on finding freely usable images.

Announcing Devices

The Bible recounts that during the time of King Saul, the people of Israel were at a significant technological disadvantage compared to the Philistines. The military capacity of the Philistines was overwhelming. They had three thousand chariots, six thousand horsemen in their cavalry, and as many soldiers in their army as there are grains of sand on the seashore. Interestingly, the number of soldiers is expressed in the same way as God's promise to Abraham regarding his descendants, almost as if to express that God was blessing the enemy more than his own people. You can find this story in 1 Samuel 13.

Verses 19 to 22 read:

> Not a blacksmith could be found in the whole land of Israel, because the Philistines had said, "Otherwise the Hebrews will make swords or spears!" So all Israel went down to the Philistines to have their plow points, mattocks, axes and sickles sharpened. The price was two-thirds of a shekel for sharpening plow points and mattocks, and a third of a shekel for sharpening forks and axes and for repointing goads. So on the day of the battle not a soldier with Saul and Jonathan had a sword or spear in his hand; only Saul and his son Jonathan had them.

The Philistines had a monopoly on blacksmithing, which allowed them to control their neighbors. Israel depended on the Philistines for their survival because plowshares, mattocks, axes, and sickles were the tools they needed to prepare and till the land, harvest grain for making bread, and cut firewood for cooking and heating.

When Goliath the giant challenged the people (1 Samuel 17), they were rightfully terrified and intimidated. He was a reliable representative of Philistine dominance. The Bible describes him not only in terms of his stature but also for the military technology he possessed:

> A champion named Goliath, who was from Gath, came out of the Philistine camp. His height was six cubits and a span. He had a bronze helmet on his head and wore a coat of scale armor of bronze weighing five thousand shekels; on his legs he wore bronze greaves, and a bronze javelin was slung on his back. His spear shaft was like a weaver's rod, and its iron point weighed six hundred shekels. His shield bearer went ahead of him (1 Samuel 17:4-7).

David was just a young boy with no military experience when he arrived at the camp to bring food to his brothers. He was not even there to fight, but in an act of courage, he confronted the giant. You know the story. David brought down the giant with a single stone that struck him in the head. When the giant fell face down, David did not take his victory for granted, but instead:

> David ran and stood over him. He took hold of the Philistine's sword and drew it from the sheath. After he killed him, he cut off his head with the sword (1 Samuel 17:51).

Now three people in Israel had a sword, and the biggest one belonged to David. This was not only a subversive act and a great conquest, but also a metaphor for transformation. In the following years, David succeeded Saul as king and "stole" the technology from the Philistines in such a way that during his generation, Israel became a regional power. David democratized Iron Age technology so that his people could enjoy great prosperity.

Towards the end of his reign, before passing the scepter to his son, he had amassed large deposits of gold, silver, bronze, and iron, ensuring the construction of the temple in Jerusalem, and safeguarding the future of the nation:

> I have taken great pains to provide for the temple of the Lord a hundred thousand talents of gold, a million talents of silver, quantities of bronze and iron too great to be weighed, and wood and stone. And you may add to them (1 Chronicles 22:14).

Think: Modern technology has long been in the hands of today's Philistines, and it is time for the church to take the sword from the giant and use it to reach those who need to hear the gospel and to bless the people of God. Clearly define the swords you need to wield to fulfill the colossal commission. Here are some questions to help you think with your team:

What changes do we need to make to our website? How can we use organic and paid search to reach people?

How will we use which social media platforms? What content plan will we implement and in what timeframe? How will we interact with private messages and chat?

Which types of content are most appropriate for the audience we want to reach? What do we need to produce that content? What devices will we use: computers, cameras, software?

What model or method of personal evangelism will we employ to mobilize believers for evangelization?

What will our live streaming consist of? How can we improve the quality of image and sound? What microphones, lighting, and set design will we use?

What traditional media outlets will we employ, and what is needed for each of them: radio, television, cinema, or print? Online? What about less traditional ones like smart TVs, voice-activated personal assistants, or music platforms?

We live in new times. It is time to mobilize the mobile device to present the message of salvation in a personal way. Adopt a simple application for each person to share the gospel with audiovisual material. Prepare your congregation, guide them to install the application, and go out together to the streets. Moreover, you can create groups on social media or have a form on your website to collect salvation stories and testimonies.

It is time to mobilize the mobile device.

Arrange: Once you have a clear understanding of the media and tools to be used, put on your producer hat. A producer oversees and manages the creation and production of a project, whatever it may be. The producer is responsible for ensuring that the project is completed on time, within budget, and to the desired quality. They are also responsible for hiring the creative and technical team, whether paid or volunteer, as well as negotiating agreements with sponsors and distributors.

Implement: Just as I have done for the previous two mission objectives, it is a good idea to use a device map for your outreach technology. **Figure 25** on the next page provides a guide for you. Use this guide to plan the budget, security, and long-term maintenance of all your devices.

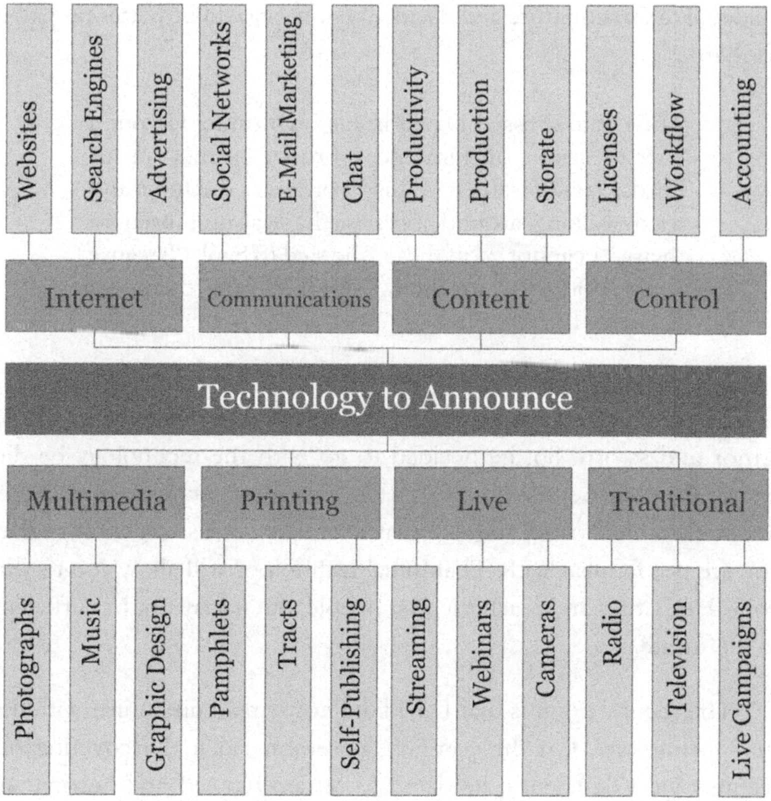

Figure 25: Technology to Announce Categories

Lead: Follow the example of David. Although he knew he would not be able to build the temple in Jerusalem, he made provision for his successor to complete all the work immediately after he stepped down from the throne.

Managing the devices for outreach involves thinking about the continuity of their use and making any necessary adjustments along the way to ensure that the work does not stop and that the technologies that best fulfill the mission are adopted.

I have two more things to share about the story of David. The first is that when David was about to fight against Goliath, all he had with him was his shepherd's staff and his sling. Saul saw him with this inadequate warrior attire and decided that he should be better prepared for battle:

> Then Saul dressed David in his own tunic. He put a coat of armor on him and a bronze helmet on his head. David fastened on his sword over the tunic and tried walking around, because he was not used to them. "I cannot go in these," he said to Saul, "because I am not used to them." So he took them off (1 Samuel 17:38-39).

David was "not used to them," referring to the technology of the armor and sword. So, he decided to go with the technology he did know, the one he was familiar with. Any new technology will feel uncomfortable at first, and you may even have to use technology that you are not familiar with. That initial technology will allow you to win your first battles and reach the first people, but it may not be sufficient in the future.

The second point is that David did not remain unfamiliar with the new technology. On the contrary, he embraced it. Discovering his technological deficiency, he dared to retrain himself and become an expert. You can see this throughout the rest of his story as king, but to illustrate the point, let us jump to a time when he was fleeing from Saul, who sought to kill him (1 Samuel 21:8-9):

> David asked Ahimelek, "Don't you have a spear or a sword here? I haven't brought my sword or any other weapon, because the king's mission was urgent." The priest replied, "The sword of Goliath the Philistine, whom you killed in the Valley of Elah, is here; it is wrapped in a cloth behind the ephod. If you want it, take it; there is no sword here but that one." David said, "There is none like it; give it to me."

In his haste to flee, David had left behind his sword and other weapons, but he was reunited with Goliath's sword, "the best" that Ahimelek could offer him. Over time, David had become an expert swordsman; he had recognized his deficiency in that first encounter with the giant but had matured in the use of technology.

In my book *Cybermaturity*, I present a seven-step map of technological maturity that ranges from initiation to innovation, serving as a management model for growing with ministry technologies. [78]

In summary, technology for outreach:

- Provides powerful means of communication to fulfill the mission at all levels.
- Enables us to expand the reach and influence of our ministries.
- Removes barriers and facilitates reaching the ends of the earth.
- Amplifies our message through multiple media and traditional channels.
- Attracts volunteers and puts tools for sharing the gospel in the palm of their hands.
- Assists in raising financial resources for missions and evangelistic activities.
- Prepares the metaversal church to evangelize cyber subcultures.

And thus, we reach the end of **Chapter 8**. You can make technology an ally to reach the world and fulfill the Great Commission to go, proclaim, and disciple. You can do the same with technology to assist, as we will see in the next chapter.

TECHNOLOGY TO ASSIST

Love your neighbor as yourself.

– GOD, LEVITICUS 19:18.

Do not say to your neighbor,
"Come back tomorrow and I'll give it to you"—
when you already have it with you.

– KING SOLOMON, PROVERBS 3:28.

Mister Tom was over eighty years old when he decided that he wanted to learn how to use a computer. Taking advantage of a state vocational rehabilitation program, he joined the computer school where I trained people with difficulties in basic tools.

Every day was a challenge with Mr. Tom. He understood the concepts very well and could easily apply them, but he struggled a lot with interacting with the computer through the mouse. Clicking never clicked for him. He could not see the correlation between the mouse movement and the cursor movement on the screen. It was extremely difficult for him to stop the mouse movement when the cursor was over a specific area, let alone clicking on an icon. Double-clicking? Forget it!

He ended up clicking on everything except what he wanted, opening screens that were not supposed to be opened or closing documents without saving them by mistake. He worked very hard to learn how to use the mouse. Every class was very frustrating for him as he made little progress.

I wish I could report that after six months, Mr. Tom learned how to use the mouse, but sadly, he never did. Nevertheless, he continued attending the class and persevered every day because he said, "At least here, they are helping me every day."

Helping those in need is a task for today and every day. Assisting others is realized through acts of love, for which technology helps you to help others.

Technology helps you to help others.

The church can assist others and achieve greater results in their mercy ministries by leveraging today's technology in multiple ways. This is the premise of this chapter.

Compassionate Towards One Another

The Bible is full of references urging us to care for and protect the orphan, the widow, the poor, the stranger, and the oppressed. During that time, these individuals were identified as the ones in greatest need of assistance and most deserving of mercy.

God Himself expresses His commitment to take action to care for the poor, the oppressed, and the needy:

> "Because the poor are plundered and the needy groan, I will now arise," says the Lord. "I will protect them from those who malign them." (Psalms 12:5).

As an expression of His love and character:

> He defends the cause of the fatherless and the widow, and loves the foreigner residing among you, giving them food and clothing (Deuteronomy 10:18).

As an extension of His love, He commands us to care for the orphaned and abandoned, just as He does:

> Defend the weak and the fatherless; uphold the cause of the poor and the oppressed. (Psalms 82:3).

It is not an option but a dimension of our mission and spirituality that is frequently reminded to us so that we do not forget it. The people of Israel had forgotten it, and by neglecting the Lord's command, they received the reproof from the prophet's pen:

> Is not this the kind of fasting I have chosen: to loose the chains of injustice and untie the cords of the yoke, to set the oppressed free and break every yoke? Is it not to share your food with the hungry and to provide the poor wanderer with shelter— when you see the naked, to clothe them, and not to turn away from your own flesh and blood? Then your light will break forth like the dawn, and your healing will quickly appear; then your righteousness will go before you, and the glory of the Lord will be your rear guard. Then you will call, and the Lord will answer; you will cry for help, and he will say: "Here am I." If you do away with the yoke of oppression, with the pointing finger and malicious talk. (Isaiah 58:6-9).

Notice the negative terms used to describe technology in this passage: chains and yoke straps that represent slavery. Helping others includes freeing those who have been subjected to the negative effects of technology, rescuing those addicted to video games, social media, gambling, pornography, or compulsive online shopping. But if technology binds, technology also rescues.

Now, take note of the promise. When we attend to the needy with technology, bread, shelter, and clothing, we will receive clarity of purpose and be healed. When we seek justice for others, we will be treated with justice. When we lift the "undesirable" in society, the glory of God will lift us up. And when we pray, God will answer us.

The New Testament echoes these passages when the apostle writes to us about the true meaning of religion:

> Religion that God our Father accepts as pure and faultless is this: to look after orphans and widows in their distress and to keep oneself from being polluted by the world (James 1:27).

God is interested in people of low resources or status in society and commands us to have the same concern.

> Be kind and compassionate to one another (Ephesians 4:32a).

Starting with our own family:

> Anyone who does not provide for their relatives, and especially for their own household, has denied the faith and is worse than an unbeliever (1 Timothy 5:8).

This mutuality of compassion and kindness is an expression of God's nature and a call for us to act in love. It is the foundation for the common good. Although technology can be used for negative and harmful purposes, technology for helping others is intended to practice compassion and bring freedom in various fundamental aspects, including within the family, among others:

1. It provides up-to-date technological tools to connect the needy with the resources they lack. We become bridges of blessing.
2. It helps coordinate assistance resources in the community through communication, tracking, and analyzing the impact of your ministry.
3. It encourages service and participation from the body of volunteers, affirming the priesthood of all believers.
4. It establishes a strong and visible presence in your sphere of influence. It answers the question: If this ministry were not present in this community, would they miss us?

The Good Samaritan

On one occasion, an expert in Jewish scriptures approached Jesus to test Him with a cunning or malicious request:

– "Tell me, what must I do to inherit eternal life step-by-step?"

Since Jesus had an extraordinary ability to perceive people's intentions, He turned the question back to the expert:

– "Well, you're the expert. What do the Scriptures say?"

In a demonstration of skill, the man combined two of the most important passages from Moses' writings: the *Shema* from Deuteronomy 6:5 and the Golden Rule from Leviticus 19:18. He responded,

– "Love the Lord your God with all your heart, with all your soul, with all your strength, and with all your mind," and "Love your neighbor as yourself."

– "Good answer!" Jesus replied. "Do this and you will live."

You can read the story and compare it for yourself in Luke 10:25-28, Mark 12:28-32, and Matthew 22:35-40. All three Gospel writers recount the event, but Luke expands its meaning with a narrative that has become known as the parable of the Good Samaritan.

In the parable, a certain man was traveling down the infamous road from Jerusalem to Jericho, known as the "bloody path," when a gang of robbers, common in the region, attacked him, stealing everything he had, stripping him, and beating him until he was left half-dead.

Three different men saw the dying man by the side of the road. Two of them were religious figures, a priest, and a Levite, who, upon seeing him, passed by on the other side and did not help him. The third was an ordinary man. He was not religious. He was not even Jewish. He was a Samaritan—a despised resident of Shechem, whom the Jews of that time did not even consider to be human:

> Two nations (*éthnesi*) my soul detests, and the third is not even a people (*éthne*): Those who live in Seir (*samareías*), and the Philistines, and the foolish people that live in Shechem. [79]

Yet, it was this despised and foolish Samaritan who felt compassion for the assaulted Jew and helped him. He approached, treated his wounds with wine and oil, bandaged them, lifted him onto his own animal, took him to an inn, and even paid the innkeeper in advance to take care of him in his absence.

What does this story have to do with our discussion? Well, I do not believe that the Samaritan was a nurse or that he carried a first aid kit with him. Instead, with basic knowledge of health, he used the three technologies available to him to help the person in need: both wine and oil are products of technological manufacturing processes, and he certainly tore a piece of clothing to use it as bandages, and clothing is also a technology.

Notice that two of the technologies he used as they were manufactured: wine and oil, but for one of them, he had to make modifications and adapt it to the specific situation: the bandages.

Your ministry of compassion today can do the same. Make use of the available technologies to help the needy, the overlooked, the abandoned, the mistreated, the marginalized, the immigrant, the forgotten, and the dying. Some technologies you can use as they come, while others you will have to modify to fit the need. The processes of adapting technology are also technology themselves.

Some technologies you can use as they come, while others you will have to modify.

Technology in mission does not exclude the love for neighbors as an option but integrates it as part of the mission itself.

At the end of the parable, Jesus asked the expert in the scriptures who was the neighbor, and he responded that it was the Samaritan. We often think that the neighbor is the one in need; we think that helping the neighbor means helping the needy. But the Samaritan was the helper, not the one being helped. The neighbor is the one who shows mercy with technology. In other words, you are the neighbor. In Jesus' own words (Luke 10:37):

– "Go and do likewise."

That is where we should also go. If we are good Samaritans, we go as imitators of God because God is our helper:

> The Lord is my helper; I will not be afraid. What can mere mortals do to me? (Hebrews 13:6).

Assisting with Technology

The story of the Good Samaritan and Jesus are powerfully connected. Jesus, though not of our "kind," became like one of us to reach our condition, which was half-dead. He was not of this world but came to this world. He was a foreigner. Strange. Undesirable. The prophet Isaiah tells us that "he had no beauty or majesty to attract us to him, nothing in his appearance that we should desire him" (Isaiah 53:2). He received the curses and insults that were meant for us. He had compassion for us, dedicating his talent, time, and treasure. He promised to return and assigned another (the Holy Spirit) to care for us in his absence. And he promised to pay for all our care until our full restoration. Jesus Christ is the greatest Samaritan of all.

Therefore, loving our neighbor represents Jesus Christ. Love fulfills the commandment: "Go and do likewise." Love is action. We live in a culture where love has been reduced to a feeling, a desirable and romantic thought. But love in the Bible is not a feeling; it is intentional and participatory action. We dedicate ourselves to helping, and we can do it with technology.

When we fulfill this pillar of the Christian mission, following the example of the Good Samaritan, we love our neighbor:

1. *We put ourselves in the place of the afflicted*: The Samaritan felt compassion. Use the feeling of empathy, compassion, to put yourself in the other person's shoes and be moved to mercy. Mercy comes from Latin and means to put the heart (cordis) in the misery (miser) of others. Technology can help you discover such misery, to investigate and understand the issues that afflict those you need to help. In the era of the Internet, we all pass by someone abandoned on the sidewalk of despair, so we have no excuse.

2. *We give what we have*: The Samaritan shared his talent, time, and treasure. Use your knowledge and skills to act boldly and

effectively. Stop on the road and dedicate time to the need you see. Use the technology you have at hand (wine, oil, donkey, hotel), modify the technology that does not perfectly address the need (bandages), or create new technology if necessary. Use technology to manage the finances of your initiative, from fundraising to the distribution of aid.

3. *We involve others*: The Samaritan recruited the innkeeper to care for the injured man while he returned. Compare Luke 10:34 and 10:35. In verse 34, it says, "he took care of him," and in verse 35, he transfers the responsibility to the innkeeper: "take care of him." Recruit. Technology serves to enroll volunteers in your cause, manage their service, educate them, and maintain continuous communication during their participation. It also helps you track the resources of your aid, such as food from a food bank, shelter in a refuge, or the comprehensive health of a medical clinic. All logistics relies on technology.

4. *We ensure recovery*: The Samaritan made sure the man would recover. This is an area where some compassion ministries fail: They give fish but do not teach how to fish. It is essential to use technology to follow up with individuals enrolled in our assistance programs, going beyond the temporary resolution of pressing needs and facilitating the process of social integration and restoration to self-sustaining life. Use technology to rediscover, reconnect, replenish, refuge, and reeducate. Furthermore, how will you know if you are making the expected progress? You can rely on tracking technologies and data analysis to assess the impact of the ministry.

> *Use technology to rediscover, reconnect, replenish, refuge, and reeducate.*

TAIL Winds

Regularly, we serve as a family at the church food bank. It is a remarkable experience. People must make appointments to attend, arrive on time, register their arrival to receive their shopping voucher, take a cart, and stroll through the aisles organized like a high-end supermarket.

Seeing the faces of families as they shop without money for fresh vegetables, meats and fish, frozen foods, cereals, dairy products, eggs, and even bakery desserts is priceless. The atmosphere is electric. We do not just provide food; we provide salvation and hope. Many have stayed in the church because of the compassion and dignified treatment they receive. Several join the food bank to serve others in need.

It is impossible to run a food bank without the logistical technology that drives the operations smoothly. From transportation, sorting, and inventory management to case management software, technology is present at every step.

Assisting People

When people finish their shopping, a group of volunteers packs the groceries while asking if they need prayer. If they say yes, the prayer team prays for them on the side and then shares testimonies of healing and provision that occur right there. Before leaving, another team asks if they would like to learn a second language or join a class to learn how to write a resume or search for employment; classes that meet simultaneously in adjacent rooms. Once the visit is completed, another group takes their bags to the car and blesses them on their way home.

Think: Who are your pastors? Perhaps you are not actively involved because you are waiting for the Levites and priests to take care of the ministry, and you need to rethink who the Good Samaritans are. What work groups do you need to form? What are the roles of each position? How will you manage communication and assign them to their respective positions so that each person knows where they belong? Think with pen and paper, or even better, with a screen and keyboard.

Arrange: Given my experience in software, I am always stationed at the check-in desk to enter people into the system and assign them their supermarket voucher. But, like me, every person in the program knows their station and what is expected of them. To ensure that everything runs smoothly, there is a lot of organization involved. I cannot provide a detailed list of every service area that exists. You must do that for your specific ministry. However, as an example, a food bank like this needs to organize:

1. Supplier relationships.
2. Provision transportation.
3. Inventory control.
4. Cleaning staff.
5. Stocking the aisles.
6. Volunteer management.
7. Community relations.
8. Marketing and promotion.
9. Transportation for the disabled.
10. Prayer group.
11. Evangelism group.
12. Facilitators for classes.

That gives you an idea. You can create your own list by thinking systematically.

Organize your volunteers. It is important to offer clear, varied, and sufficient opportunities for people to serve. To do this, promote service opportunities, recruit volunteers, and follow up with them. Collaboration and project management software tools can help you coordinate activities and efficiently manage the work of volunteers.

Implement: Assignments, locations, confirmations, and all communication are handled with a dedicated tool. Upon arrival, you go to a volunteer check-in station, search for your data on a tablet, and print a name tag that tells you where you will be serving that day. Plan every possible task that people need to do according to the activity calendar.

Coordinate with other mission areas for community outreach activities to promote your ministry. For example, a park fair can integrate worship, outreach, and assistance to serve others.

Lead: Guide everyone to the same standard of quality, both paid staff and volunteers. I mean that sometimes we treat volunteers with different demands and rigors, but the commitment to successful programs should be the same for everyone. Without a commitment contract, you will experience efficiency leaks. The Word reminds us of this in Colossians 3:23-24:

> Whatever you do, work at it with all your heart, as working for the Lord, not for human masters, since you know that you will receive an inheritance from the Lord as a reward. It is the Lord Christ you are serving.

To do this, consider the processes and personnel management system. There is much to learn from professional models in this area. As a specialist in HR software implementation, I have created the management diagram shown in **Figure 26** to guide you in managing the personnel of your outreach programs.

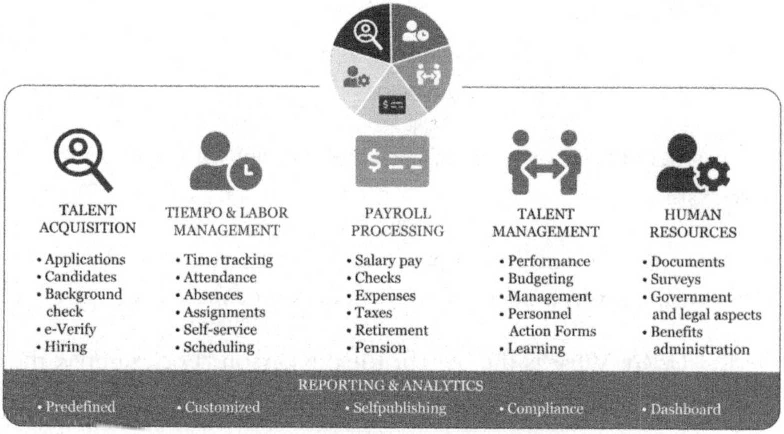

Figure 26: Human Resources Management Diagram

Assisting Systems

One way to organize activities related to assistance systems is to think in terms of the user experience. User experience is a technical term that denotes the feeling a user experiences when using your services. The idea is to put yourself in the shoes of the person receiving the assistance and walk through the entire process as if you were the recipient, paying special attention to each interconnected system. Use this concept as the driving force behind your assistance systems.

Think: Consider the example of a food bank. We can relate a separate but interdependent system for each group of people mentioned in the previous section. Thus, we would have a system for:

1. Engaging with suppliers.
2. Transporting provisions to the distribution center.
3. Managing inventory.
4. Maintaining and cleaning the warehouse.
5. Stocking the aisles with food.
6. Managing volunteers.
7. Building relationships with city authorities.
8. Marketing and promoting the program.
9. Providing transportation for people with disabilities.

10. Praying for needs.
11. Evangelizing those receiving assistance.
12. Educating the community.

Any beneficiary of your services will interact with one or more of these systems.

Arrange: To deliver a superior experience, you should consider the following seven aspects of the beneficiary's experience:

1. *Utility*: What is the benefit for the person? For example, they will receive food to feed their family.

2. *Desirability*: What is the quality of the benefit? For example, high-quality nutritious food that is not expired.

3. *Accessibility*: How easy is it to obtain the benefit? For example, no need to pay or fill out complex forms.

4. *Credibility*: How consistent and reliable is the assistance? For example, it is received at the agreed-upon place and time without fail.

5. *Findability*: How easy is it to locate? For example, the address is accurate and as advertised.

6. *Usability*: How easy is it to interact with the systems to schedule an appointment and collect the groceries?

7. *Value*: What additional value do the systems offer? For example, while picking up groceries, they can also learn English.

Implement: Plan the step-by-step journey of your beneficiaries. They may not be called beneficiaries, but the idea is the same. For example, a person in need of food receives a brochure from the food bank, schedules an appointment through the online application to visit the center, is greeted by a welcome team, is assigned a case agent to enter them into the system and provide them with a purchase voucher, takes a shopping cart, and goes through the aisles. In the aisles, they find the food they need because the inventory manager stocked them beforehand with provisions brought by the transporter, thanks to the public relations efforts with suppliers.

Plan the step-by-step journey of your beneficiaries.

Lead: Once you have planned the user experience, ensure that you provide excellent service in interaction with the systems, from the well-communicated brochure on how to get started to the prayer team praying and the dispatching team taking the food to the car. Each person in the system should reflect the values and message of the Gospel in their most genuine expression of love.

Manage multiple channels of community service, such as afterschool tutoring, legal assistance, mobile health clinics, or government-sponsored programs. Use search engines to discover available assistance and connect your volunteers with government programs or other Christian compassion organizations.

You can also create your own directory of professional services with an up-to-date online database for everyone to consult.

Assisting Processes

The church's food bank has been very successful, to the point that three locations, two mobile distributions, and a health clinic have been opened. The work we do has reached the ears of city authorities and even the federal government, to the extent that it has received several official recognitions. Its light is shining.

Recently, I heard something remarkable and unexpected from its founder. He said, "My vision is for there to be no food banks in the city." In other words, the goal is to eradicate the need. Although Jesus Christ said that we would always have the poor with us (Matthew 26:11), the idea of a future where every family in the city can be self-sufficient is the transformative story that assistance processes seek.

Think: In the words of renowned author Stephen Covey, "begin with the end in mind."[80] If your idea is to eradicate hunger in your community or empower self-sufficient families, think about how you will achieve it. Perhaps you should start by understanding the current condition of families to intervene positively and measure your progress.

The future of the food bank is about changing lives and destinies, breaking the cycle of poverty through dignity and honor, relationships and partnership, creativity and beauty, and generosity. These are defined themes to outline the processes of how we deliver assistance at the food bank.

Think about how technology can help you identify the needs of your community to engage in outreach and relief ministry. For example, collect, visualize, and analyze demographic and psychographic data, behavioral trends, and prevailing patterns of needs. Make use of digital maps that provide application programming interfaces (APIs) to visualize your own data.

Arrange: It is essential to use technology responsibly and ethically to fulfill the ministry's objectives and bring God's love to those in need. The following are examples of technological processes that can be employed in a ministry of helping others.

1. *Focus on social media*: Stay relevant and leverage the multiplier effect of your posts. Use surveys, tags, and hashtags to discover needs and challenges. Share news, events, and specific assistance needs. Use social media to connect with people in need and provide emotional support or direct them

to available resources. I remember when I visited the corporate offices of Focus on the Family in Colorado. They had set up a 24-hour monitoring room to identify individuals with suicidal intentions, present them with the Gospel, and refer them to specialized authorities and other resources.

2. *Manage projects*: Monitor the progress of the project from start to finish. With specialized tools, you can coordinate volunteer efforts and ensure that ministry objectives are being met.

3. *Open microfinancing channels*: Track the money process from donations to bank accounts. Crowdfunding websites can be effective ways to raise funds for community assistance initiatives.

4. *Create mobile applications*: Mobile applications published on digital markets and stores can be an effective way to reach your community. They can include forms, surveys, and other tools to interact with those already connected to your ministry, or they can distribute timely assistance information using mobile push notifications. They are also very efficient in communicating important information to volunteers.

5. *Establish follow-up processes*: Monitor progress and evaluate the impact of your ministry in helping others. Identify areas for improvement and closely track the growth and transformation of your users.

Implement: Develop plans based on your key themes and chosen initiatives, considering the time, people, and resources required. A social media promotion campaign or the creation of a mobile application are perfect candidates for creating an action plan that unfolds over time, with clear and achievable goals. If you do not know the methodology for managing a project, seek information and educate yourself.

Lead: In process administration, it is important to document and have a clear and detailed record of how things are carried out, subjecting them to scrutiny to identify areas for improvement and adjust them accordingly. Technology can help automate processes to make them more efficient and reduce human errors. There are also tools that allow you to regularly monitor processes to ensure they are being carried out effectively and that desired objectives are being achieved. Remember that it is essential to train both staff and volunteers to ensure they understand the processes and can carry them out effectively and efficiently.

Assisting Policies

Working with the community has great rewards, such as seeing the faces of dignified and hopeful people, or the joy of children receiving immediate healing or parents responding to the Gospel of Jesus Christ and joining the church. But it also comes with great responsibilities and risks.

Sometimes things do not go as planned. There is always someone who complains or causes trouble when faced with certain limitations, or someone who wants to take advantage of the nobility of mercy. That is why it is of utmost importance to have clear and well-communicated policies, rules, and procedures through all possible channels to prevent misunderstandings and undesirable behaviors.

Think: What rules should staff and volunteers follow? What are the policies that govern community service? There may even be some rules from the city or the government. What are the procedures to ensure that everything is done properly, with order and excellence? These are the types of questions you should ask in ministry meetings, with a focus on both pragmatism and procedures, and considering the supporting technology.

Arrange: In the time of the apostles, "The Church was made up of Aramaic-speaking Jews from Palestine and Greek-speaking Jews from the Diaspora." [81] The racial and cultural differences between these groups had begun to create tensions within the ministry. Generally, the Palestinian Jews considered themselves superior or purer than the diaspora Jews, believing that the latter had mixed racially and adopted worldly practices. This disdain created a disparity in how Hellenistic Jews were treated. We read about this in Acts chapter six:

> In those days when the number of disciples was increasing, the Hellenistic Jews[a] among them complained against the Hebraic Jews because their widows were being overlooked in the daily distribution of food (Acts 6:1).

We have already discussed how Scripture urges us to take care of widows, so the complaint was legitimate. Given the circumstances, the apostles called for consensus to organize the equitable distribution of food.

> So the Twelve gathered all the disciples together and said, "It would not be right for us to neglect the ministry of the word of God in order to wait on tables. Brothers and sisters, choose seven men from among you who are known to be full of the Spirit and wisdom. We will turn this responsibility over to them and will give our attention to prayer and the ministry of the word." This proposal pleased the whole group (Acts 6:2-5a).

They chose Stephen, a man full of faith and of the Holy Spirit; also, Philip, Procorus, Nicanor, Timon, Parmenas, and Nicolas from Antioch, a convert to Judaism. 6 They presented these men to the apostles, who prayed and laid their hands on them.

> They chose Stephen, a man full of faith and of the Holy Spirit; also Philip, Procorus, Nicanor, Timon, Parmenas, and Nicolas from Antioch, a convert to Judaism. They presented these men to the apostles,

> who prayed and laid their hands on them (Acts 6:5b-6).

The list of names is significant because it can be inferred that representatives from both sides were chosen to attend to everyone equally. No additional complaints appear in the rest of the book of Acts, which allows us to infer that they resolved the situation. Furthermore, the same passage confirms:

> So the word of God spread. The number of disciples in Jerusalem increased rapidly, and a large number of priests became obedient to the faith (Acts 6:7).

Establishing proper order is a biblical practice. It brings peace, common well-being, and full freedom for the exercise of ministry.

Organize your policies to become an accessible ministry. Do not exclude the disabled, people with special needs, or those in cultural disadvantage from the ministry of the church. Instead, facilitate their participation from the core of your organization. For example, you can make your website accessible by following the guidelines of the World Wide Web Consortium (W3C),[82] or you can translate its content to several languages with the use of multilingual modules.

Implement: In the food bank, there are established policies to ensure order and equitable distribution of quality food to beneficiaries. Here are some examples:

1. There are set schedules and rules for food distribution, and we ensure that beneficiaries are aware of the established policies and procedures.

2. We employ a tracking system to keep a record of those who receive food and to prevent duplicate deliveries. Such a system is required for reporting to donors and relevant authorities.

3. We allocate a specific amount of food based on the household size and needs, considering factors such as families with infants who may require formula and diapers.

4. We define clear admission criteria, including income requirements and an assessment of economic needs, to determine eligibility for assistance.

5. We have clear procedures for food distribution in cases of scarcity or excessive demand, ensuring that resources are distributed equitably. This was particularly crucial during pandemic lockdowns.

6. We regularly review key metrics to measure the impact of our ministry and ensure that established goals are being met.

7. We follow safety measures to protect, store, and handle food. Additionally, we provide a safe and secure environment during visits.

Write and organize policies that are suitable for your ministry in your specific context.

Establishing proper order is a biblical practice.

Lead: Every act of mercy and compassion requires good practices in policies and procedures regarding the use of technology. While there may not have been any ill intent in the preferential treatment given to the widows of Palestinian Jews, better planning corrected the disparities.

In the words of John Stott:

> It is not suggested that the oversight was deliberate... more probably the cause was poor administration or supervision. [83]

Poor administration and supervision can be the cause of many undesirable consequences, which is why controlled management carried out by qualified individuals is crucial. This includes the proper and transparent handling of finances. Let me explain.

The call of the seven was to "serve the tables." The tables were not only used for serving food or eating but also for exchanging money and completing commercial transactions. Remember that "Jesus entered the temple courts and drove out all who were buying and selling there. He overturned the tables of the money changers and the benches of those selling doves" (Matthew 21:12). What Jesus did is a topic for another book, but you can see the use of tables for financial transactions and appreciate the importance of managing finances with order and justice.

Assisting Devices

Throughout this chapter, I have identified several devices of assistance, both in the biblical passages and in the references to the food bank. When we attend to the needs of the needy and the vulnerable, Isaiah 58:10 promises us that our light will shine brightly:

> And if you spend yourselves in behalf of the hungry and satisfy the needs of the oppressed, then your light will rise in the darkness, and your night will become like the noonday.

Light is a natural phenomenon; it is not technology, but certainly, there is technology that serves to make light shine. A lamp on a shelf illuminates all those who are in the house. That is what Jesus said:

> You are the light of the world. A town built on a hill cannot be hidden. Neither do people light a lamp and put it under a bowl. Instead they put it on its stand, and it gives light to everyone in the house. In the same way, let your light shine before others, that they may see your good deeds and glorify your Father in heaven (Matthew 5:14-16).

You will need different devices and tools to make your assistance ministries shine, depending on the size and specific needs of the program. It is important to assess the needs of "everyone in the house" before acquiring devices and starting their operation.

If there are foreigners who speak other languages, you may need to offer classes in the dominant language as a second language and a closed-circuit radio or online applications for live interpretations. Details can be found in **Chapter 11** under "Translation."

Think: Reflect with your ministry team on the technologies that bring light to your community, in other words, those that you need to use in your compassion and mercy programs.

Create a map of your initiatives and the necessary technology: What kind of community action programs do you have in the church, and what technological resources could enhance their effectiveness? What are the main challenges you face, and how can you overcome them with technology?

Consider the information you need to collect and how you will maintain and use it: What kind of information do you need to gather and analyze to measure the impact of your compassion programs, and what technological tools can assist you in achieving those goals?

Consider the technical capacity of your team: What are the technological skills of your team, and what kind of training do you need to effectively use the necessary devices and software? What kind of maintenance do these devices or software applications require, and how will you ensure that they are up-to-date and safe to operate?

Finally, secure the finances for each project: What budget do you have to invest in technology, and how can you maximize your spending to achieve the greatest benefit?

Arrange: Technological knowledge facilitates better organization to serve others, using the right tool for each job. In the same passage from the book of Matthew, Jesus says that we are the salt of the earth. The passage reads:

> You are the salt of the earth. But if the salt loses its saltiness, how can it be made salty again? It is no longer good for anything, except to be thrown out and trampled underfoot (Matthew 5:13).

Several translations of this passage contain the interpretation that it refers to common salt, or sodium chloride, used to preserve and season food, which is why it is written: "How will it regain its flavor?" However, the same context of the passage gives us a clue that it is not about "table salt" but about "the salt of the earth."

Table salt does not become tasteless or degrade naturally, while the salt of the earth does lose its quality. The salt of the earth is another kind of salt, more likely phosphate salt, which was used for two specific purposes.

The first purpose was to place it on the floor of clay ovens to regulate their temperature. It is well known that salt has thermoregulatory properties and could be used to heat the ovens faster and make more efficient use of fuel. That's why I prefer the NIV translation, that uses saltiness. Saltiness is about its qualities instead of how it tastes:

> You are the salt of the earth. But if the salt loses its saltiness, how can it be made salty again? It is no

longer good for anything, except to be thrown out and trampled underfoot. (Matthew 5:13).

The second use was as a fertilizer for agricultural land, as phosphate is a catalyst for agricultural processes to accelerate ripening and promote natural seed production. This phosphate is gradually released into the soil until its nutritive properties fade away and it must be discarded by the roadside.

You see, each salt for each use. And that means organizing, so arrange everything with structured order for its proper use. The possibilities of using technology to assist are endless.

Utilize technology to coordinate logistics using applications designed for case management or CMS (Case Management System). This type of software provides your ministry with a central repository to store and track information about program beneficiaries, including contacts, types of assistance, progress notes, appointments, and the history of the mutual relationship.

Arrange everything with structured order for its proper use.

Implement: Having decided which community action programs you are going to initiate or continue, you can now plan the implementation of relevant tools and devices, as well as their recurring use.

I invite you to create a hierarchy of the assistance devices that your programs need to use and keep it visible and up to date. The previous **Figure 27** can serve as a guide in creating your own hierarchy of devices and software.

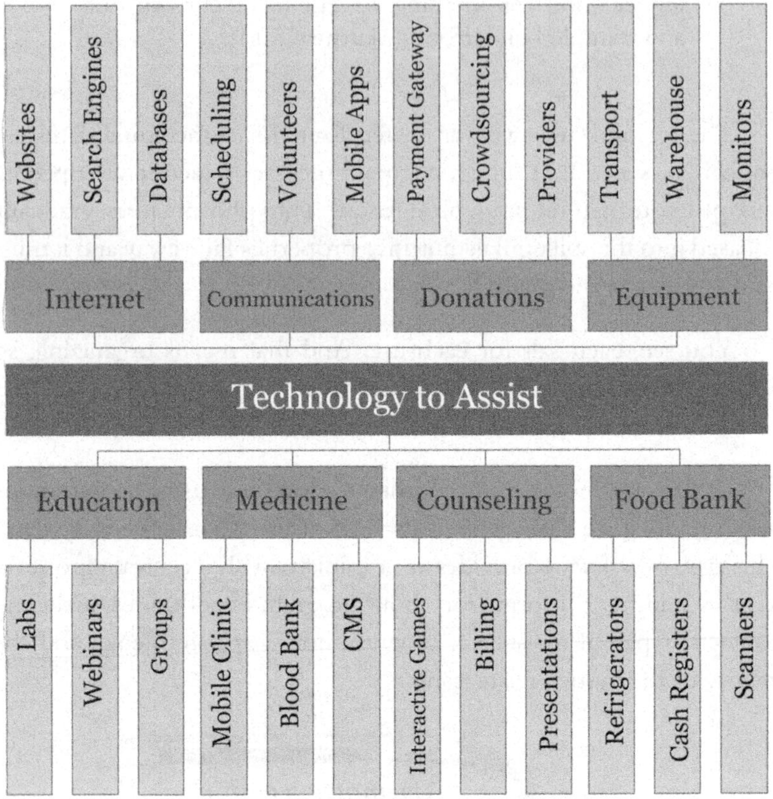

Figure 27: Technology to Assist Categories

Lead: Now, manage how you select and properly use your devices and software. To do this, you need to consider the following aspects:

1. *Maintenance*: Regular maintenance of equipment and software is necessary to prevent failures and ensure proper functioning.

2. *Updates*: Keeping systems and applications up to date is important to improve security and functionality.

3. *Training*: Personnel responsible for using the equipment and software need to be trained to ensure proper usage and avoid technical and operational issues.

4. *Security*: Implementing security measures is important to protect user data and ministry information.

5. *Access Controls*: Establishing an access control system is necessary to prevent unauthorized use of equipment and software.

6. *Usage Policies*: We have already covered this. It is important to establish usage policies for personnel responsible for using the equipment and software to ensure proper usage and avoid technical and operational issues.

7. *Usage Logs*: Keeping a record of equipment and software usage is important to monitor their usage and detect possible issues.

8. *Budget*: Establishing a budget for equipment renewal and maintenance is necessary.

These general aspects also apply to any other mission areas we have discussed. This list applies to all devices in the ministerial technologies' framework. I recommend you go back and review the devices to adore, assimilate, announce, and associate in light of this list.

Let us review. Technology to assist aids you in the exercise of mercy and compassion because it:

- Drives you into love for your neighbor.
- Communicates your service goals and values.
- Connects solutions with the problems that afflict people.
- Facilitates the administration and detailed control of your resources and beneficiaries.
- Makes you present in your community and opens opportunities to share the Gospel.
- Tracks all users in all your programs.

- Recruits and manages volunteers serving in the various opportunities you offer.
- Manages the finances of your initiatives.
- Measures and visualizes the impact you are creating and helps in decision-making.

That concludes **Chapter 9**. The mission is not the lamp or the salt, not even the light or the heat, but to reflect the glory of God, manifest His love, and provoke His praise. Let us move on to consider the final pillar of technology in mission: technology to associate.

TECHNOLOGY TO ASSOCIATE

Technology is nothing. What's important is that you have faith in people.

– STEVE JOBS, COFOUNDER OF APPLE.

Two are better than one, because they have a good return for their labor: If either of them falls down, one can help the other up. But pity anyone who falls and has no one to help them up.

– KING SOLOMON, ECCLESIASTES 4:9-10.

The doctors had given Patricia only three more months to live unless she received immediate treatment. Her heart failure was worsening every day, and the pacemaker that had sustained her for the past ten years was reaching its last beat. There was only a small window of time to extend her life by replacing the pacemaker. The rest would be a waiting game, hoping that her heart would have enough strength to withstand the procedure.

Patricia had another problem. She lived in a Latin American country with limited resources and little purchasing power. The bad news came when the country was under a strict lockdown due to the pandemic, limiting her mobility and her ability to arrange for emergency surgery. Furthermore, the pacemaker alone cost fourteen thousand dollars. The prospects were not encouraging.

But Patricia never lost hope. She had been connected to one of the online communities of the church. She would sometimes lead the worship time in the live streams. Everyone appreciated her charisma and her faithfulness to God expressed through service and acts of mercy. Now it was her turn to be on the receiving end.

As soon as she shared her situation in the community chat, we sprang into action to help her, seasoned with much prayer and supplication, with thanksgiving, reminding her of Philippians 4:6-7:

> Do not be anxious about anything, but in every situation, by prayer and petition, with thanksgiving, present your requests to God. And the peace of God, which transcends all understanding, will guard your hearts and your minds in Christ Jesus.

How appropriate. The peace of God would guard her heart! We asked her about the resources she had available and what she could do to raise some of the money. She had started a crowdfunding webpage that was already yielding its first fruits. We obtained the clinic's estimate for the procedure and the anticipated costs, which we used to spread the word, and an organization joined in to collaborate, negotiating to lower the total cost. Then we reached an agreement with a nonprofit organization to serve as an intermediary for collecting the most significant donations.

Some generous donors completed the efforts. The pacemaker was imported. The surgery date was set, and on the day of the operation, the entire community was in an international prayer chain with great expectations.

What would have been the outcome of this story without technology? Patricia would have found herself isolated, unable to receive much support from the local church due to its closure and unable to share her need with the faith community. But Patricia was not alone. The internet, chat groups, video conferences, email, online commerce, and crowdfunding all came together to accompany her on her journey.

David Gomez, an assistant minister of the Assemblies of God in Spain, aptly writes in his book *Evangelizing on the Internet*:

> The internet is the fastest-growing medium of communication, and it is so because it is much more than just a communication medium; it is a medium of connection and relationship. [84]

Technology strengthens our communications to strengthen our relationships. With technology, we can accompany people in their journey with God, addressing their needs, nurturing their growth, fostering their hope, and even igniting their hearts. Technology enables us to live in community.

Technology strengthens our communications to strengthen our relationships.

As for Patricia, everything turned out well. Her heart beats to the rhythm of the praises she still plays and sings. It is worth mentioning that her heart beats thanks to the medical technology that accompanies her everywhere. Accompanying each other with technology brings life and is transformative.

Koinonía pneumatos

It saddens me to think that there may be those who profess the Christian faith personally but have ceased to believe in the Church. It is possible that we hold misconceptions about what the Church means, thinking that it refers only to historical traditions or official institutions, but that would be a chronic reductionism. The Church is the dynamic community of Christian life. We must not forget that believing in the Church is part of our confession of faith.

To that effect, the Nicene-Constantinopolitan Creed of AD 381, accepted by the Church at large, reads:

> And I believe in the Holy Ghost, the Lord and Giver of Life; who proceeds from the Father and the Son; who with the Father and the Son together is worshipped and glorified; who spoke by the prophets. And I believe one holy catholic and apostolic Church. I acknowledge one baptism for the remission of sins; and I look for the resurrection of the dead, and the life of the world to come. Amen. [85]

This is the last section of the Nicene Creed which pertains to the Holy Spirit. I frequently like to say that although the Triune God participates in the entire plan of salvation, they have been individually expressed, as:

1. Creation is the visible work of God the Father.
2. The incarnation, death, and resurrection are the visible works of the Word, God the Son.
3. The Church is the visible work of God the Holy Spirit.

To not believe in the Church would be to deny the visible work of the Holy Spirit in the history of humanity. The Holy Spirit is the power (*dynamis*) from heaven that empowers believers to testify:

> But you will receive power when the Holy Spirit comes on you; and you will be my witnesses in Jerusalem, and in all Judea and Samaria, and to the ends of the earth (Acts 1:8).

The Holy Spirit incorporates us into the community of faith through confession and baptism:

> For we were all baptized by one Spirit so as to form one body—whether Jews or Gentiles, slave or free—and we were all given the one Spirit to drink (1 Corinthians 12:13).

The Holy Spirit is the life-giving power of the Church. Its fruit is the fruit of the Spirit, and its gifts are the gifts of the Spirit. The Holy Spirit is also the power of life that raised Jesus Christ from the dead and guarantees, seals, and gives us a foretaste of experiencing the same resurrection:

> The Spirit gives life (John 6:63a).

> And you also were included in Christ when you heard the message of truth, the gospel of your salvation. When you believed, you were marked in him with a seal, the promised Holy Spirit, who is a deposit guaranteeing our inheritance until the redemption of those who are God's possession—to the praise of his glory (Ephesians 1:13-14).

> And if the Spirit of him who raised Jesus from the dead is living in you, he who raised Christ from the dead will also give life to your mortal bodies because of his Spirit who lives in you (Romans 8:11).

Thus, the Church, being one, holy, catholic, and apostolic (refer to the section on the mission of the Church in **Chapter 5**, starting from page 122), is the fellowship of the Spirit or in the original Greek, *koinonía pneumatos.*

We find the combination of these two words in the farewell of the second letter to the Corinthians and the letter to the Philippians:

> May the grace of the Lord Jesus Christ, and the love of God, and the **fellowship of the Holy Spirit** be with you all. (2 Corinthians 13:14 - emphasis added)

> Therefore if you have any encouragement from being united with Christ, if any comfort from his love, if any common **sharing in the Spirit**, if any tenderness and compassion, then make my joy complete by being like-minded, having the same love, being one in spirit and of one mind (Philippians 2:1-2 - emphasis added)

Fellowship in the Spirit produces unity of mind, unity of love, unity of soul, and unity of thought in the Church. A community is a common unity. Unity, not uniformity. The fellowship in unity is a faculty of being filled with the Spirit and can only be fully realized in the Church.

The meaning of the word *koinonía* is very rich. On one hand, it means fellowship, communion, participation, and fraternity. It denotes the mutual commitment of a two-way relationship. It is derived from the word *koinos*, which means something common, shared, or of common agreement, in contrast to the word *idios*, from which "idiosyncrasy" is derived, and which means private or individual.

The Christian life is not meant to be experienced in a private and individualistic manner, although these are ideals in the prevailing culture. No! The Christian life is not idiosyncratic but *koinosyncratic*. The Christian life is designed to be lived in the company of the community of believers; it is shared life in God:

> Above all, *koinonía* describes the shared life in Christ... This shared life has its source in God and is much more than participating in a part or portion of a whole (like sharing a pizza or owning shares in a corporation). *Koinonía* is participating in the whole, being an integral part of a totality. [86]

On the other hand, *koinonía* also means to conduct a business in partnership, to be co-owners of a property or company, or the administrative management of running an association. It implies that being in partnership requires the proper management of the partnership itself. The contract of *koinonía* demands that in business, all parties share a common vision, mission, values, strategies, and objectives, and each puts forth the necessary focus and effort to fulfill those purposes. *Koinonía*, more than just a casual friendship, is a mutual commitment to shared goals and achievements, where everyone contributes to the whole to reap the benefits and profits.

In this way, in *koinonía*, we accompany each other by being collectively responsible for the logistics of our journey together. It is like embarking on a trip where everyone has tasks and is responsible for their success —some buying tickets, others arranging accommodations, others managing the itinerary, and others ensuring food and entertainment. *Koinonía* is both companionship and the administration that comes with it.

Administer Your Associations

Once upon a time, there was a manager who had a reputation for being a poor administrator. He was not filled with *koinonía pneumatos*. When his boss reviewed his job performance and held him accountable, the manager knew that his job was at risk, and it was likely that he would be fired soon. So, he devised a dishonest but effective plan to win favor with the clients. He thought, "If I get fired, at least one of the clients will hire me, and I won't be left out on the streets."

We accompany each other by being collectively responsible for the logistics of our journey together.

He did not care that this attitude violated the company's conflict of interest policies. So, he picked up the phone, called the two most important buyers, and altered their invoices of what they had to pay. He gave the first one a 50% discount and the second one a 20% discount.

When the boss called the manager for the second time, he thought it would be the end of his employment. But well, the damage had already been done. The boss had found out about the unjustified and exaggerated discounts the manager had given without authorization. To his surprise, the boss named him Employee of the Month instead of firing him. It turned out that the clients were so delighted that they posted five-star reviews on a popular public review site, praising the company's excellent customer service. They also left viral comments on social media, and organic search visits to the company's website increased. Unexpectedly, sales had a significant increase in less than a week.

This story, or well, something similar, was told by Jesus Christ and can be found in Luke 16:1-15. Verse 8 reads:

> The master commended the dishonest manager because he had acted shrewdly. For the people of this world are more shrewd in dealing with their own kind than are the people of the light (Luke 16:8).

The supposed mismanagement of the manager had resulted in good customer engagement. The moral of the story: if you are astute in business, people will accompany you. The counter-moral: for more people to accompany us, people of faith must manage better than those who do not know faith.

The sad reality that we must change is that many ministries have not developed the same astuteness in managing the father's business as those who conduct worldly business have. It is time to shift the balance and be wiser in managing ministry resources.

Associating with Technology

An astute manager of the Father's business must be responsible and diligent. They must manage with wisdom to advance the kingdom of God and accompany those in the family of faith. An astute administrator should be faithful in managing the community and able to address challenges and problems that may arise. They should also have a mindset of stewardship rather than ownership, recognizing that everything we have and manage belongs to God, and we should use it according to His will and purpose. Ultimately, an astute manager does everything for the glory of God and the fulfillment of His purposes.

Technology for accompanying can be our best ally in astute administration and walking alongside others in the pilgrimage of faith, regardless of barriers of time and space. Pastor Angela Craig, the pastor of the first digital church of the Assemblies of God, Pursuit Church Live, writes:

> Today, technology can help us create deeper connections and community while ministering to those who might otherwise remain beyond our reach. [87]

No wonder worldly businesses are astute in the use of technology. Those of the light, then, can mediate *koinonía* through technology.

TAIL Winds

I am a founding member of our church's online community program in Venezuela. The idea of having communities in exile connected with the ministry of the central church had been on the pastor's mind for a while, but the necessary actions had not been taken to initiate it. The lockdowns during the pandemic were the catalyst that sparked the first conversations to create such a ministry.

Initially, we had been informally connecting with a community of friends in Miami, Florida, for several weeks. After the pastor's call to initiate the process of formalizing online communities, I invited the leader of the community to be part of the leadership team and for his community to serve as the pilot community for the program, to which he also agreed.

When writing these lines, the program has six communities. In addition to the pilot community, communities have been added in San Jose, Costa Rica; Lima, Peru; Santiago, Chile; Buenos Aires, Argentina; and Rome, Italy. The program is thriving, growing, and flexible enough to explore the necessary adaptations as we return to normal life.

Associating People

How did we start? First, we organized the core leadership team of five key individuals, now known as the global team, who meet week after week to imagine, pray together, and shape and bring life to the program. With some ideas on paper, a lot of enthusiasm, and support from the leader's experience in Miami, we created a reproducible model and invited five properly identified leaders to take charge of each of the communities.

Think: Who are your apostles today? Who are your digital apostles? The word "apostle" is a combination of two Greek words: *apó*, a directional preposition meaning outward, and the verb *stello*, which means to send or dispatch. Jesus first sent the apostles to do the work of ministry, who in turn passed the baton to the next generation of believers. We see this transition of leadership in Paul's words to his disciple Timothy:

> And the things you have heard me say in the presence of many witnesses entrust to reliable people who will also be qualified to teach others (2 Timothy 2:2).

If you want to accompany your community, assign trustworthy and qualified individuals to teach others through their words, but above all through their example.

Create a team of people dedicated to data entry. This is what we did in the church to feed the visitor and new believer information into the church management system after services. Do everything necessary to keep all the information centralized in one system, as this facilitates its management in the medium and long term.

Any information system is only as good as the information it contains. As the Bible says, "A man reaps what he sows" (Galatians 6:7b). On the other hand, accurate information will allow you to create precise reports and analyze the true state of your ministry.

Depending on the size of your congregation, you may need specialized groups for each area. Managing accounting systems is different from handling personal data of new converts, students, or their parents.

Arrange: Structure the team model, lines of responsibility, and functions of each member. Document everything. Have a routine of meetings to report on progress and challenges along the way. Take advantage of tools like the ones mentioned earlier to stay organized.

It is important to create volunteer groups for leaders and servers and document them in your preferred word processor. Consider using an application that is easily accessible to everyone in the international communities.

For online communities, we created a reproducible model with seven circles of support, each with well-defined roles, which I include here as they may serve as a guide for you.

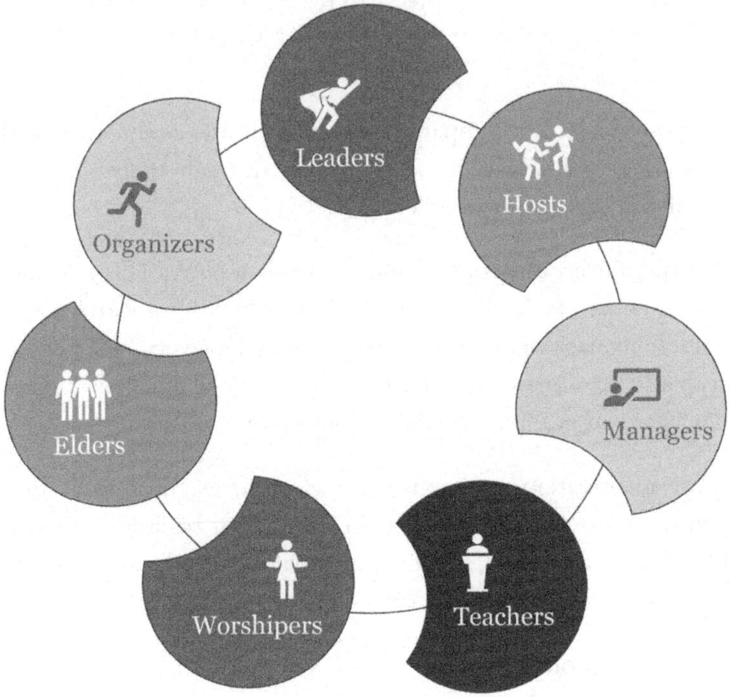

Figure 28: Association Circles

Each circle in **Figure 28** above has the following responsibilities assigned:

1. *Organizers*: Responsible for initial formation, logistics, and coordination of activities.

2. *Leaders*: In charge of the spiritual health and growth of the online community; think of them as digital pastors.

3. *Hosts*: Deal with the technical aspects during online meetings to ensure people can connect.

4. *Community Managers*: Responsible for keeping the community alive during and after meetings.

5. *Teachers*: A tested group responsible for teaching Scripture during meetings and through pre-recorded videos.

6. *Worshipers*: Musicians and worship leaders who lead praise during meetings and through pre-recorded videos.

7. *Covering Elders*: Spiritual elders responsible for providing spiritual coverage to the community.

Implement: Your team of leaders and volunteers should be capable of accompanying others in their faith journey. Some important qualities for carrying out this task effectively are:

1. *Maturity*: It is important for those who accompany others in their faith to be spiritually mature and have a solid knowledge of Scripture. This will enable them to share the advice of God's Word in everyday life.

2. *Empathy and compassion*: Leaders should be able to put themselves in other people's shoes and understand their needs and difficulties to offer appropriate support and guidance.

3. *Active listening*: It is crucial for leaders to listen attentively to individuals without judgment or criticism to better understand them.

4. *Patience*: Each person's faith journey is unique and may take time and effort. Leaders should be patient and respect the pace of growth for everyone.

5. *Follow-up*: After each contact, provide follow-up and continue supporting individuals in their spiritual journey.

Plan not only the activities and tasks but also the personal growth in character and qualities of those who accompany others.

Lead: Maintain responsible and transparent management of the ministry's resources and ensure they are used effectively for the benefit of the faith community, whether digital or in person.

While the online community program is running, manage the creation of promotional materials for meetings, chat groups to encourage participation between meetings, confirmations with the teacher of the day, recording meetings, and uploading them to the designated application, among other necessary actions. Each assigned person mobilizes to carry out these tasks on time.

Communicate clearly to all involved parties which communication channels they should use and be consistent with the content, frequency, and objectives of your communication.

Associating Systems

Define the support systems that need to be coordinated for people to feel connected to your ministry and their needs addressed for growth. If multiple systems are involved, define their roles and interactions with each other. Consider how the different components of technology-mediated support come together. Use the TAIL model to define these systems and their mutual dependencies.

Think: First decide on the support methods that best fulfill your purposes of creating and strengthening your community.

Thanks to the development of communication technologies, the possibilities are vast. For example, you can have online or instant messaging meetings, develop mobile applications to promote Bible reading, meditation, and prayer, or use social media to share testimonies. There are even worship and music applications that can facilitate an online worship experience, as well as financial management systems for keeping track of contributions. Some of these systems have been socialized to facilitate interaction with others, incorporating elements of social networking, gamification, and support groups.

Organize: Your systems should function smoothly to avoid misunderstandings and setbacks. Frequently consult the expectations of the individuals within your ministry to be assertive in meeting their needs. In online communities, we conduct surveys regularly to identify material, prayer, growth, and learning needs and propose creative ways to address them. If you use a specialized survey system, you can analyze historical data to make decisions using the same survey.

Allow your systems to promote participation and feedback from those being accompanied to improve your programs and foster an attitude of respect, empathy, and compassion for all. Rely on effective communication and follow-up models to achieve these objectives.

Implement: Plan a period of trial and error that allows you to adjust your systems along the way. Be flexible. When we started online communities, we tested various chat applications before settling on one. For capturing emails and sending automated sequences, we started with one provider, but its limitations led us to switch to another. This other provider would also help us in creating landing pages to promote regular meetings and special conferences.

Plan the initiation, selection, assimilation, implementation, use, and evaluation of your systems, as each stage will require your full attention and dedication.

Lead: Everything that is managed can be measured, and everything that is measured can be improved. This means that you need to manage everything that can be measured: finances, attendance, volunteer participation, student enrollment, employee skills, and the list goes on. The best way to manage the resources of the ministry is by using a specialized software system.

I recommend choosing a Church Management System (ChMS). These systems offer a combination of modules and options to manage different areas of the ministry. They typically consist of various specialized modules that integrate with each other (the *koinonía* of systems, but that is a topic for another day). Implement one module at a time to see immediate and medium-term results.

Manage the support systems according to their capabilities and technical requirements. Although the principles are the same, there are considerable differences between managing your live streams and managing online communities. Online streaming is a less participatory process and, if it could be interactive, it would be less comprehensive since the only form of interaction is through chat. Online communities assume the fact that you can see the person in real-time – if they turn on their camera, of course – allowing for more direct and participatory interaction.

Manage the time of your meetings and leave enough space for participation. Allocate a few minutes for natural interaction in the virtual lobby and allow members to ask questions and make comments. Open small sessions for discussion and prayer.

Manage the use of systems on a shared calendar. The regularity of online community meetings suggests a usage cadence. In our case, we chose a video call system with options to automatically schedule meetings by integrating with our preferred calendar.

We created a shared calendar for sending short videos with micro-messages and birthday greetings, as well as for addressing prayer requests. We also created an email calendar for teachings. All these systems working together have helped our online communities thrive and grow.

Associating Processes

Faith accompaniment processes can be mediated by technology. New technologies can enable quality and personalized spiritual guidance regardless of geographical location or time constraints, allowing more people to be accompanied in their faith journey and training a new generation of gospel-committed ministers.

Although we may not think about it frequently, accompanying faith with technology has been a reality since the early days of the Church. With the expansion of the Church, early Christians saw the need to narrate the gospel consistently in every place. The first gospel to be written was that of Mark, but it was later followed by the books of Matthew, Luke, and John. Luke, being a companion of Paul on his missionary journeys, also undertook the task of narrating the Acts of the Holy Spirit among the nations, sometimes working as Paul's amanuensis in some of his letters.

The gospels and the writings of Paul and Luke would soon become encyclicals (from the Greek *enkyklios epistolè*), letters addressed to various communities to be read by them and passed on to the next community in the circuit. Often, these communities made a handwritten copy before circulating them, which is why we have fragments and abundant variants of the same portions of the New Testament.

You can see the process of technology in action to accompany the faith of Christians in the known world. Today's technologies serve to accelerate, personalize, specialize, translate, and contextualize faith more efficiently, if they are used intentionally for that purpose.

Think: Every system adopted will bring about a transformation in the way of work. It is important to recognize that any transformation takes time, but it is necessary to be intentional about the habits you want to change.

Think about what processes you can employ to accompany others in their faith. These processes have two aspects: processes related to sharing relevant faith in community and processes related to managing faith communities.

1. Processes for Sharing Relevant Faith in Present Circumstances. Here, you define processes to:

    a. Identify the needs of individuals participating in the programs.
    b. Select and utilize appropriate tools, such as video conferencing applications, digital reading, and online communication.
    c. Assign follow-up agents, mentors, and counselors to guide people in their faith.
    d. Schedule regular and special support sessions.
    e. Prepare supportive and learning materials, such as reading resources and audiovisual materials.

2. Processes for Managing Faith Communities: Managing faith communities is as important as the accompanying processes themselves. For this, you detail processes such as:

    a. Develop clarity in the vision and key objectives of the communities.
    b. Maintain a record of each community, all its members, and their attendance at activities.
    c. Manage logistics, volunteers, and growth activities.
    d. Develop and maintain budgets for community initiatives, stewardship, and donations.
    e. Regularly evaluate growth and seek new opportunities to expand the ministry.

Arrange: Each process will require specific organizational considerations, starting with defining specific goals, training individuals involved in each area, managing the flow of information, allocating sufficient resources, and ongoing evaluation to ensure program continuity. I have seen several online community programs frustrated by the post-pandemic return to the physical church and a lack of clarity in sustaining online communities in the long term. Adjust your programs to the changing realities of your environment. Have a forward-thinking and progressive mindset, organize with an expectation of expansion and continuity.

Organize with an expectation of expansion and continuity.

Implement: Project the reflection and organization of your processes with a focus on accompanying believers in your faith communities in their faith, confession, and growth.

The celebration of sacraments in online communities, for example, has created unique challenges and opportunities. How can you celebrate the baptism of a member of the online community who wants to publicly profess their faith and be incorporated into the body of Christ? How can you celebrate the Lord's Supper from a distance? And what about the dedication of children? These require different approaches and alternative plans, such as partnering with churches in different cities to celebrate baptism or using food distribution services to deliver the elements of the Lord's Supper prior to the celebration – I have not seen any church doing this yet, but it is possible.

Lead: Leading involves regularly reviewing and adjusting the other areas of TAIL as needed. This means reviewing and adjusting what you think, what you organize, and what you plan for each accompanying process. Additionally, you must direct and control the processes themselves and the mechanisms for review and adjustment in a feedback cycle. It is advisable to set a regular period for review and adjustment that aligns with each specific process, for example, every six months or every year for online communities, or every year or every two years for the ministerial formation of new believers. Dedicate dates in your calendar for these reviews.

Associating Policies

There is a biblical body of policies that servants and administrators must follow in the work of ministry. When Paul writes to Timothy regarding the qualities of church leaders, he details the conduct requirements for deacons and their wives:

> In the same way, deacons are to be worthy of respect, sincere, not indulging in much wine, and not pursuing dishonest gain. They must keep hold of the deep truths of the faith with a clear conscience. They must first be tested; and then if there is nothing against them, let them serve as deacons. In the same way, the women[b] are to be worthy of respect, not malicious talkers but temperate and trustworthy in everything. A deacon must be faithful to his wife and must manage his children and his household well. Those who have served well gain an excellent standing and great assurance in their faith in Christ Jesus (1 Timothy 3:8-13).

The text functions almost like a checklist, and if there is a deficiency in any of the points, it is considered sufficient reason to temporarily disqualify the person from the diaconate. Deacons are responsible for the health of relationships within the faith community, the faithfulness of teaching, and good administration. That is why they must have healthy relationships in their homes with their wives and children, have a clear conscience to uphold the truths of the faith, and not engage in dishonest business practices when handling money. Deacons must adhere to the highest standards of conduct to represent well the faith in their community.

Similarly, we are responsible for defining the policies and procedures that are to govern the ministry in its function of accompanying others in the faith.

Think: What policies protect your faith community? Are there child protection requirements? Which policies facilitate growth processes, and which are essential in the environment where you minister? Which policies are mandatory for using the devices and software applications that you have defined as tools for accompaniment? Do you have membership requirements? More critically, what are the financial and accounting requirements for the legal representation of your ministry? What about the obligations to provide accurate reports to denominational bodies and government entities?

Here is a brief list of ideas to outline your ministry policies:

1. Use of temple facilities.
2. Acceptable use of digital resources such as software, websites, and social media.
3. Privacy of community members' data.
4. Use, maintenance, and replacement of devices in service of the community.
5. Policies for child and elder protection.
6. Requirements for leaders, pastors, elders, and deacons.
7. Requirements for community members.
8. Requirements for civil association or legal representation.
9. Policies for financial management and taxes.
10. Emergency and disaster preparedness.
11. Policies for interorganizational relationships, such as councils, denominations, and other voluntary associations.

All these things are part of the policies of accompaniment. List them and be ready to organize, plan around them, and manage them with transparency and loyalty. For each of these policies, there are associated technologies, both for their execution and for their administration. Include the description and use of the corresponding technologies with each policy as appropriate.

Arrange: Once you have defined the body of policies and procedures that will allow you to better accompany your faith community in person and through digital means, it is necessary to organize them coherently.

When organizing these policies and procedures, consider the following:

1. *Clarity*: Ensure that policies and procedures are written clearly and are easily understandable for all members of your faith community.

2. *Approval*: Ensure that policies and procedures are approved by the church hierarchy to avoid conflicts and ensure compliance.

3. *Implementation*: Ensure that policies and procedures are effectively implemented to achieve the established objectives.

4. *Communication*: Ensure that they are communicated to everyone, so they are understood and applied correctly.

5. *Evaluation*: Ensure that they are periodically evaluated to adjust to changing needs.

6. *Training*: Ensure that all members of the faith community are trained to effectively implement the policies and procedures.

7. *Accessibility*: Ensure that they are accessible to all members of the faith community, including those with various disabilities.

8. *Compliance*: Ensure that they are consistently followed to guarantee a safe and welcoming environment for all members of the faith community.

Implement: Plan the collection of relevant information regarding the requirements in your context, conduct preliminary analysis, develop each policy and procedure, allocate time for implementation, and ensure that your community operates smoothly with them.

Also, plan where each of your policies will be published. Some will be for public use, such as your policies for child protection that require you to review the criminal background of individuals who work with children or file income taxes, while others will be for private use, such as the guidelines for maintaining membership in good standing with your denomination and the quarterly reports you must submit to the board of directors. By the way, if you have reporting requirements, create a calendar to work on them with sufficient advance notice.

Lead: Remember that administration is a gift of the Holy Spirit, as we shared in the section on administration in **Chapter 5**. This means that the Holy Spirit gives us the ability to administer and lead God's flock, following biblical principles of good stewardship.

Administer the policies of accompaniment with the fullness of the Holy Spirit. Ensure that the policies are properly implemented, and that staff and volunteers are aware of them. Stay at the forefront of the training and supervision process and be the first to apply the policies to yourself. Become an advocate for the policies, with the ability to explain them clearly to the congregation and address any questions and concerns that may arise with love.

Review and update the policies and procedures regularly to ensure they remain relevant and effective in accompanying a faith community that is constantly changing. Administer change and how it impacts your policies.

Associating Devices

The title of this section says it all: most human beings are accompanied by devices. For example, we have a global population of 7.4 billion people, and there are 7.7 billion active smartphones with subscriptions. This means that many individuals have more than one device.

In fact, smartphones are compact multi-purpose computers, and as such, they can accompany people in their faith journey in many ways, some of which we have already discussed. It is worth considering that the foundation of mobile companionship lies in mobile applications.

Administer change.

Think: When it comes to accompanying others with technology, there is technology on both sides of the communication process. In other words, there is transmission technology and reception technology. This is important because for every device, gadget, or software application you want to use, there are elements on both sides.

Do you want to stream your services live? You need to consider the devices and software for transmission, and you need to ensure that the devices and software used by those connecting can effectively receive the transmission. Sometimes, you may even need to provide end-user support to ensure the success of your transmission. Do you want to regularly connect with your community through your own mobile application? You need to develop the application and publish it on your preferred digital store, as well as provide users with the necessary guidance to download and learn how to use it. When you think about the devices for accompaniment, consider the requirements on both sides of the equation.

Arrange: When King Solomon set out to begin the construction of the temple in Jerusalem, he reached an agreement with King Hiram I of Tyre to provide him with all the wood for the project. The narrative can be found in 2 Chronicles 2.

Solomon sent a letter to Hiram requesting technical and material assistance for the largest worship technology project in history. Solomon wanted to hire an expert in multiple technologies from Tyre (people first):

> Send me, therefore, a man skilled to work in gold and silver, bronze and iron, and in purple, crimson and blue yarn, and experienced in the art of engraving, to work in Judah and Jerusalem with my skilled workers, whom my father David provided (2 Chronicles 2:7).

He also requested the importation of cedar, cypress, and algum wood from Lebanon:

> Send me also cedar, juniper and algum logs from Lebanon, for I know that your servants are skilled in cutting timber there. My servants will work with yours to provide me with plenty of lumber, because the temple I build must be large and magnificent (2 Chronicles 2:8-9).

In return, Solomon would provide Hiram with wheat, barley, and oil. The favorable commercial exchange for both kingdoms would result in praise to God. Hiram would accompany Solomon as the project's main partner, devising a maritime system to efficiently transport the wood from Lebanon to Jerusalem, a journey that represented about 250 miles by land:

> We will cut all the logs from Lebanon that you need and will float them as rafts by sea down to Joppa. You

> can then take them up to Jerusalem (2 Chronicles 2:16).

This plan required technology on both sides of the equation. Hiram would be responsible for cutting and transporting the wood by sea to Joppa. Solomon would then take it from Joppa to Jerusalem (about 40 miles) and shape it for the construction of the temple. When you design a technology program for accompaniment, you need to organize the usage on both sides of the participants.

Implement: Another thing we can observe from this passage is the level of planning and coordination necessary to carry out such a grand project. Solomon was a great project manager with the wisdom given to him by God. That is why he wrote:

> Commit to the LORD whatever you do, and he will establish your plans (Proverbs 16:3).

For the construction of his royal palace and the temple, he had thousands of people under his command, "he conscripted 70,000 men as carriers and 80,000 as stonecutters in the hills and 3,600 as foremen over them" (2 Chronicles 2:2). I cannot even imagine the level of planning. I have managed many technology projects in my career, but never with more than twenty people at a time.

For online communities, a mobile application was created that allows community members to stay informed about upcoming activities, access recorded teachings, read the Bible, request prayer, and see when the community is praying for their needs. The best part is that it was created without knowledge of programming languages, using a no-code platform.

As you implement technology for accompaniment, it is essential to create a map of the devices you will be using in each ministry area. I provide an example in **Figure 29** below.

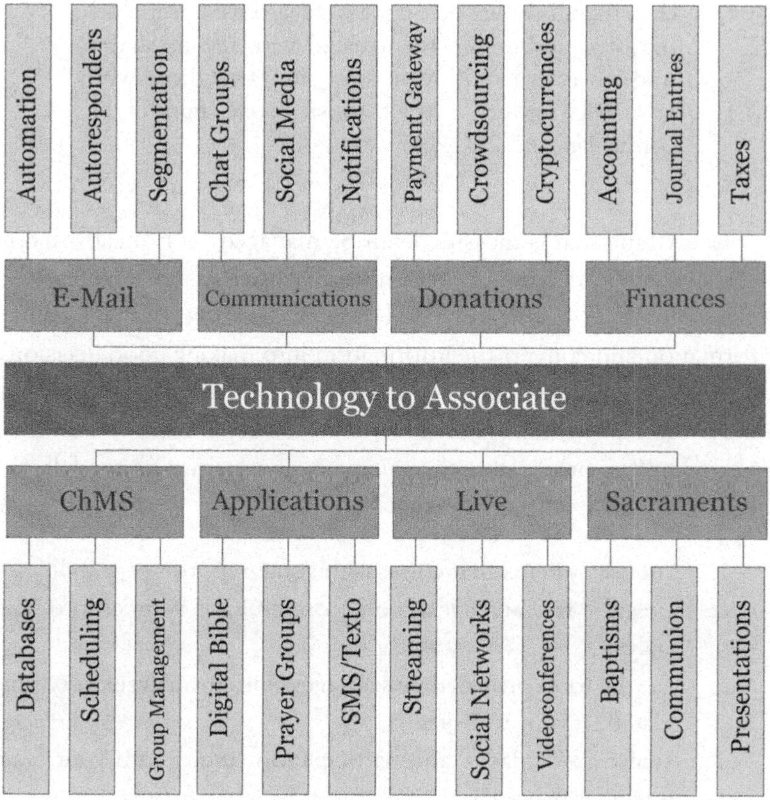

Figure 29: Technology to Associate Categories

Lead: Finally, manage with wisdom and diligence, avoiding laziness or negligence, all the resources and devices of technology to fulfill your desires to accompany people in their Christian growth, as Solomon said:

> A sluggard's appetite is never filled, but the desires of the diligent are fully satisfied (Proverbs 13:4).

Be ready to give an account of what has been entrusted to us as faithful stewards of the mysteries of God:

> This, then, is how you ought to regard us: as servants of Christ and as those entrusted with the mysteries God has revealed. Now it is required that those who have been given a trust must prove faithful (1 Corinthians 4:1-2).

Everything that is measured can be managed. This means that to be a shrewd administrator, you must become a shrewd measurer. When you have a system that is fed by data, you can turn that data into information and convert the information into making good decisions. The analysis of information is an area that is often overlooked in ministry and is the main weakness of most people I know who have already implemented a church management system. Do not fall into the same trap:

1. Choose systems that come with some reporting capabilities.
2. Integrate free analytical tools from multiple sources into your analysis.
3. Use the tools provided by search engines to analyze everything related to your website.
4. Analyze attendance and participation results with your team in regular meetings.
5. Study the ministry's financial data, donations, and the profitability and self-sufficiency of your programs.

Technology for accompaniment will allow you to add shrewdness to your ministry because, among other things:

- Increases the productivity of employees and volunteers.
- Facilitates relationships among community members.
- Channels ministry resources effectively.
- Provides clarity to ministry information.
- Supports informed decision-making.
- Amplifies attendance, participation, and donations.

By accompanying in community, you will strengthen faith. By managing shrewdly, you will receive greater rewards. Remember that if you are faithful with little, the Lord will entrust you with much, but only when you manage what you measure and measure what you manage will you know the difference between what is little and what is much.

Everything that is managed can be measured. Everything that is measured can be managed.

Technology does not replace the importance of interpersonal relationships and community in ministry, but it can help expand and enhance the way we reach people, support one another, and accompany each other on the journey of faith.

With this, we covered the entire framework for thinking, arranging, implementing, and leading ministerial technology within the mission. Each component of technology serves to minister in every area of the mission, demanding that we view technology not as a necessary evil or an afterthought, but as essential and strategic in reaching the world with the powerful and transformative message of the gospel of Jesus Christ for salvation and abundant life.

Now, some closing words.

AN APP FOR EVERYTHING

When wireless is perfectly applied the whole earth will be converted into a huge brain, which in fact it is, all things being particles of a real and rhythmic whole. [88]

– NIKOLA TESLA, SERBIAN INVENTOR.

We've all been inundated with so many ingenious, must-have, time-saving apps and tools that we really don't have a second left to spare. [89]

– RYAN HOLMES, HOOTSUITE FOUNDER.

In 2003, we were celebrating Tim's birthday, a coworker, at a restaurant in the historic center of Pasadena, California. Tim, who oversaw maintaining computers at Fuller Theological Seminary, was very excited because the grand opening of the Apple Store in the city was just a few days away.

After finishing lunch, Tim suggested that we walk to the store to see the preparations. To our surprise, as we stood at the door watching the employees make the final arrangements, Steve Jobs himself came out and invited us inside. He greeted us warmly, dressed as usual in his black turtleneck, ripped jeans, and sandals. He asked us for our opinion on how the space near the customer service counter should be organized.

Standing there, we watched him go to the back of the store, move some armchairs and tables as if it were a living room, and then come back to ask us for our opinion. He stood in front of us, looking at the changes, and said, "What do you think?" This happened at least three or four times. We were a bit shocked by his informal demeanor and how accessible he was. The last time he stood in front of us, we noticed a hole in the back of his pants, and as soon as he turned around, we made jokes about pooling our money to buy him new pants.

Unforgettable! That day, we provided free consulting to Steve Jobs.

Application Ecosystem

It would be five years from that encounter until the commercialization of the first iPhone and its app store. Apple launched its App Store in 2008 with 500 basic applications for its own operating system, iOS. However, they had created an ecosystem that allowed developers from around the world to consistently publish and distribute applications to their user base. This agility enabled them to publish over two million applications in less than ten years.

The growth of available applications was so rapid that the company launched its advertising campaign around everything that could be done with their smartphone because if you want to do anything, "there's an app for that."

For this reason, it would be futile to provide an extensive list of everything that is possible within the app ecosystem. If you also consider applications for Android, Windows, and Linux operating systems running on mobile and laptop/desktop computers, web applications, and cloud software, there is an app for everything.

Starting to implement technology with applications or devices is like taking the snake by the tail. That is why I have emphasized starting with the mindset, with strategic thinking. Ministerial technology begins with the mission and prioritizes people on both sides of the equation.

Keeping this in mind, I would like to offer you a reference in this short chapter, organizing all the technologies mentioned in this book by categories. I hope this chapter serves as a starting point and a recurring resource for you.

Applications Reference

Due to the limitations of the printed medium and the ever-changing nature of technology, I have decided to move specific references to applications and their developers to the accompanying website, with links to the respective sources of such technologies. I will keep these references updated with their ongoing development over time.

Accounting

Software to manage the ministry's accounting, finances and budgets including options for tax preparation.

https://vla.lu/account

Adoration

Presentation software, musical sequences, audio distribution, devices, equipment, and other resources.

https://vla.lu/adore

Advertising

Digital advertising platforms for search and social media and access to grants for nonprofit institutions.

https://vla.lu/advertise

Artificial Intelligence

Practical applications of artificial intelligence to create written, graphic, audio, and visual content.

https://vla.lu/prompt

Automation

Platforms for task and process automation, data flow, and transfer between programmed and automated applications.

https://vla.lu/automate

Bible Software

Bible software programs and digital Bibles for devotional reading and private and communal study.

https://vla.lu/read

Business Intelligence

Specialized software for data review, analysis, and deriving organizational intelligence for decision-making.

https://vla.lu/decide

Case Management

Applications for managing cases in the context of legal, medical, psychological, or material assistance.

https://vla.lu/assist

Church Management

Specialized software apps for church administration, group coordination, and constituent member data.

https://vla.lu/associate

Cloud

Up-to-date information on cloud computing and the resources it offers to host software applications.

https://vla.lu/host

Communications

Applications for connecting and communication channels such as email, telephony, and SMS texting.

https://vla.lu/communicate

Crowdfunding

Websites and applications for conducting fundraising and donation drive campaigns for specific projects.

https://vla.lu/finance

Data Analytics

Applications, software, and other resources for collecting and analyzing data and trends of various kinds.

https://vla.lu/analyze

Databases

Applications for storing data of various kinds in different formats, contacts, trends, and analysis.

https://vla.lu/save

Education

Applications for discipleship, managing learning, creating content, and distributing it through the Internet.

https://vla.lu/learn

Educational Content

Distribution applications for high-value and high-quality Christian content to integrate into teaching.

https://vla.lu/teach

Electronic Mail

Specialized software for creating and maintaining automated email outreach and marketing campaigns.

https://vla.lu/send

Evangelism

Mobile applications for sharing the plan of salvation, outreach ideas, and digital advertising platforms.

https://vla.lu/announce

Food Bank

Software applications aimed at managing a food bank with case studies and demographic data.

https://vla.lu/feed

Graphic Design

Applications for creating various templates and graphics for commercial use in social media and print media.

https://vla.lu/design

Identity

Applications and forms to define your corporate identity or branding to enhance the consistency of your message.

https://vla.lu/identify

Inventory

Applications dedicated to managing and maintaining physical inventory resources for the ministry.

https://vla.lu/inventory

Licensing

Licenses to acquire rights and use music, videos, photos, and other audiovisual elements with permission.

https://vla.lu/license

Live Broadcasts

Equipment and software recommendations for live streaming high-quality content on social media.

https://vla.lu/broadcast

Maps

Maps to integrate into mobile applications and websites and even interact with them programmatically.

https://vla.lu/locate

Models

Various models that leaders can use in planning, managing, and directing organizational processes.

https://vla.lu/model

No-code Platforms

No-code refers to platforms for developing applications without prior programming knowledge.

https://vla.lu/develop

Payment Gateways

Providers of digital payment and donation processing, some dedicated to the church and others with a secular focus.

https://vla.lu/give

Prayer

Mobile social applications to promote personal prayer, devotions, and follow up on testimonies in the community.

https://vla.lu/pray

Productivity

Word processing, checklists, calendars, contacts, stationary, presentations, spreadsheets, and other documents.

https://vla.lu/produce

Project Management

Specialized software for project management to coordinate delivery time, resources, and personnel.

https://vla.lu/execute

Radio and Podcasting

Instructions for creating an online radio station and publishing audio programming through simple syndication.

https://vla.lu/spread

Reputation Management

Websites for managing reputation, public opinion, reviews, and testimonials of a company or organization.

https://vla.lu/recommend

Screen Mirroring

Wireless devices and peripherals to mirror the screens of mobile devices on smart TVs.

https://vla.lu/mirror

Social Networking

Ideas for using public social networks and applications to create private social networks for the community.

https://vla.lu/connect

Storage

Document and file storage using cloud services and integrated with other software applications.

https://vla.lu/store

Surveys

Specialized applications for creating forms to conduct and distribute surveys and analyze responses.

https://vla.lu/survey

Thinking

Software tools to help us think, such as flowcharts, mind maps, and decision matrices and trees.

https://vla.lu/think

Translations

Resources and software for translating your website, video subtitles, live meetings, and all kinds of content.

https://vla.lu/translate

Video

Desktop and Internet applications for producing and editing videos, and platforms for distribution and storage.

https://vla.lu/view

Videoconference

Software for group calls, video conferences, and simultaneous collaboration with Internet connectivity.

https://vla.lu/meet

Volunteers

Applications for recruiting, selecting, managing, and tracking volunteers and their assignments.

https://vla.lu/serve

Websites and Blogs

Applications for creating websites, blogs, and modules to add usage, functionality, security, and performance.

https://vla.lu/publish

Workflow

Online applications for drawing flowcharts and shared workflow diagrams on the Internet.

https://vla.lu/diagram

Final Words

And so, we come to the end of *Cyberministry: 5 Essential Functions of Technology in the Mission of the Church.* Throughout this book, we have discussed the importance of developing a new and mindful perspective of the technological environment in which we operate in this century, examining our attitudes about technology in ministry, discovering the role of technology in some biblical stories, and considering technology as a strategic component of the church's mission.

Where appropriate, I have also identified some commonly used technologies and how they can be applied in each of the five essential mission functions. This book does not aim to be a comprehensive manual of all these technologies or a step-by-step guide. It is not even a list of all possible technologies. Rather, the book presents a thinking model, a framework, for thinking, organizing, planning, and managing technology in ministry.

As a framework, it is now your task to define the strategic purpose of technology in your ministry and outline a strategic plan that integrates technology into the core of your mission. This is no small task. On the contrary. But as the Word says:

> From everyone who has been given much, much will be demanded; and from the one who has been entrusted with much, much more will be asked. (Luke 12:48b)

Now I have entrusted you with much, so that you may make technology an allied instrument of your mission. I pray to God that you may see with new eyes, bring light to your thoughts, guide your steps, and become a purpose-driven administrator, a *kybernétes* empowered by the Holy Spirit to testify of the gospel of Jesus Christ and perfect the mission in the work of ministry.

Go forth, *cyberminister*!

APPENDIXES

APPENDIX A: TECHNO ATTITUDES

Techno Attitudes Scale

Age: _________ Gender: M F

This scale assesses attitudes towards technology in ministry. There is no right or wrong answer. What matters is knowing your perspective. Mark with an 'x' or circle 'o' the value you consider most accurate for each attitude, ranging from 0 for non-existent attitudes to 10 for strong attitudes. We appreciate your patience and accuracy.

The important thing is that you can conduct this evaluation several times throughout the year to observe how attitudes change over time and if you can identify positive or negative trends in the transformation of the technological culture of your ministry.

Name of the participant:											
Ignorance											
0	1	2	3	4	5	6	7	8	9	10	
Rejection											
0	1	2	3	4	5	6	7	8	9	10	
Blindness											
0	1	2	3	4	5	6	7	8	9	10	
Traditionalism											
0	1	2	3	4	5	6	7	8	9	10	
Delays											
0	1	2	3	4	5	6	7	8	9	10	
Indiscipline											
0	1	2	3	4	5	6	7	8	9	10	
Disorganization											
0	1	2	3	4	5	6	7	8	9	10	
Paralysis											
0	1	2	3	4	5	6	7	8	9	10	
Immorality											
0	1	2	3	4	5	6	7	8	9	10	
Infobesity											
0	1	2	3	4	5	6	7	8	9	10	
Insecurity											
0	1	2	3	4	5	6	7	8	9	10	
Excuses											
0	1	2	3	4	5	6	7	8	9	10	
Grand total											

Appendix A: Techno Attitudes Scale

Techno Attitudes Scale Totals

Tabulate the totals for each area for each participant. This will give you an idea of which attitudes are prevalent in your team and which ones you can focus on to create a suitable technology culture in your ministry.

Name of the participant	Ignorance	Rejection	Blindness	Traditionalism	Delays	Indiscipline	Disorganization	Paralysis	Immorality	Infobesity	Insecurity	Excuses	Grand Total
Total													
Averages (Total/)													

Appendix A: Techno Attitudes Scale Totals

If you would like to obtain an electronic copy of the *Techno Attitudes Scale* and all other tools for printing, please visit www.cyberministry.info.

APPENDIX B: DISCUSSION QUESTIONS

This is not a test; these are just some questions to facilitate group discussion. I have divided the questions by chapters to facilitate navigation and focus, but you do not need to use them in this way. You do not need to answer all the questions, but it is recommended that in your planning and evaluation sessions, you answer at least two or three from each section.

Chapter 1: Other Eyes

1. Why is it important to see things from a different perspective?
2. How does seeing things from a different perspective help you have a correct understanding of technology in ministry?
3. Take inventory of all the technology you can see right now and reflect on how it came to be here.
4. Can you list several technologies that make up a city?
5. What are the seven characteristics of every technology?
6. Choose a couple of those characteristics and apply them to how they are reflected in the technology you currently use in your ministry.

Chapter 2: Techno Attitudes

1. Do you consider yourself to be technophobic or technophilic? Why?
2. In what ways has your ministry resisted the use of technology?
3. Can you identify old technologies that you do use in ministry but have not seen as technologies?
4. When you think about the bronze serpent, do you consider it to be technology or not? Why?

5. What would be a balanced attitude that your ministry can adopt regarding technology?
6. List the twelve negative attitudes we can have towards technology. Which ones do you have, and what can you do to improve your attitude?
7. Is it truly possible to reject technology? Why or why not?
8. Do you believe you suffer from infobesity? If so, how can you subject yourself to an "info-diet"? Write down practical actions you can take.
9. What are the common excuses your ministry has for implementing or not implementing certain technologies? What can you do to counteract those excuses?

Chapter 3: Technology

1. What did you think about technology before reading this book?
2. In what ways has the definition of technology presented in this book changed your perception of technology? Has it changed for the better or for the worse?
3. How does your concept of technology resemble or differ from the concept presented in the book?
4. What significance does the integrated ecosystem of technology have for your ministry?
5. What significance does the cultural network nature of technology have for your ministry? What implications does this have for your specific cultural context?
6. Choose a technology in use in your ministry and identify each of the five areas in detail.
7. Who are the key individuals involved in technology in your ministry?
8. What are the components of a system?
9. In what ways is worship a system?
10. Try to diagram a ministry process following the model presented in the book.

11. What policies can you identify as necessary in the practice of your ministry?
12. What criteria can you use to choose devices and software applications?

Chapter 4: Technology Trilogy

1. What does the technology trilogy consist of?
2. How can the technology trilogy help you achieve the desired outcome of growth in your ministry?
3. What are the seven components of a strategy?
4. Do you talk about vision, but do not have it written down? What can you do to write it down?
5. Do you have your ministry strategy written down? Why or why not?
6. If you have a written ministry strategy, what are the next steps for integrating technology in each area?
7. How do you devise the commission before you commission the device?
8. Choose a technology that is already in use in your ministry, create three columns on a piece of paper, and organize the things that belong to strategy, tactics, and intelligence.
9. What tactical factors can you identify that affect the use of technology in your ministry?
10. What operational factors can you identify that affect the use of technology in your ministry?
11. Define your performance indicators, how you will measure them, and where you will record them for future review.

Chapter 5: Ministerial Technologies

1. Why is it important to have a framework and model for managing technology in ministry?
2. What does the acronym TAIL mean?
3. How does your perspective change, knowing that administering is a gift from the Holy Spirit?

4. In your own words, what is the mission of the church?
5. List the five essential functions of the mission, the 5 As, and how your ministry executes each one.
6. Is your ministry deficient in any of the five essential functions? Why? What actions can you take to remedy it?
7. What is the difference between the first chronological commandment and the first commandment in priority? Are they related? How?
8. What are ministry technologies?
9. How do ministry technologies relate to the mission of the church?
10. Draw the *Ministerial Technologies Matrix*® from memory.
11. Name at least two ways you can use the *Ministerial Technologies Matrix*® to organize and evaluate technology in your ministry.
12. How does technology serve to worship, learn, reach, help, and accompany in your ministry?

Chapter 6: Technology to Adore

1. How does Exodus 20:24 relate to technology?
2. Read Exodus 31. What does this passage have to do with technology for worship?
3. What is the relationship between music, worship, and technology?
4. What does it mean for biblical worship to be mediated?
5. What does it mean for biblical worship to be multimedia?
6. What does it mean for biblical worship to be technological?
7. How can you apply the definition of technology to the function of worship?
8. Create a SWOT analysis diagram of technology for worship in your ministry.
9. Make a detailed list of the people, systems, processes, policies, and worship devices following the TAIL model applied to your ministry.
10. Create a visual map by categories of all the worship devices you currently have or would like to have in your ministry. Use

this map to create assertive plans to enhance worship in your ministry.

Chapter 7: Technology to Assimilate

1. In what way is language essential to technology?
2. Should you apply concepts of educational technology to the discipleship processes of the church? Why?
3. How has technological development in the church affected its ability to educate in matters of faith?
4. Based on Deuteronomy 6:1-7, what does biblical instruction consist of, and why is it essential in disciple formation?
5. Why is discipleship not merely the transmission of information?
6. What are the educational tasks of the church, and how do you apply them to your ministry?
7. How can you apply the content machine to everything you want to communicate in your ministry?
8. How can you apply the definition of technology to the function of learning?
9. Create a SWOT analysis diagram of technology for learning in your ministry.
10. Make a detailed list of the people, systems, processes, policies, and learning devices following the TAIL model applied to your ministry.
11. Create a visual map by categories of all the learning devices you currently have or would like to have in your ministry. Use this map to create assertive plans to foster discipleship in your ministry.

Chapter 8: Technology to Announce

1. What is the calling of the Great Commission?
2. What is cyberculture, and what does it mean to evangelize cyberculture?

3. What is the metaversal church? What can you do to make your ministry penetrate other virtual worlds?
4. What are the advantages of technology in reaching others with the message of the gospel?
5. What areas of outreach through technology can you establish in your ministry?
6. What is the importance of knowing your community well to be more effective in your ministry, and how can technology help you with this?
7. Do you have a model for personal evangelism? How can you use mobile applications to communicate the gospel efficiently and effectively?
8. What do you still need to transmit high-quality live services?
9. How can you apply the definition of technology to the function of reaching others?
10. Create a SWOT analysis diagram of the technology for outreach in your ministry.
11. Make a detailed list of the people, systems, processes, policies, and outreach devices following the TAIL model applied to your ministry.
12. Create a visual map by categories of all the outreach devices you currently have or would like to have in your ministry. Use this map to create assertive plans to expand the outreach of your ministry.

Chapter 9: Technology to Assist

1. Why are compassion, mercy, and mutual aid essential functions of the mission?
2. How did the Good Samaritan use technology to help the half-dead man on the side of the road?
3. What are the four steps of love that we can learn from the Good Samaritan, and how can they be strengthened with technology?
4. How can your ministry utilize available technologies to help those in need, the underprivileged, and the disadvantaged?

5. Have you implemented accessibility and assistance technologies for the disabled? Why? If not, what can you do to address this situation?
6. Do you effectively manage your volunteers, from recruitment to their ministry tasks and results?
7. Have you established mutual aid partnerships with other organizations? If your answer is yes, how are you managing the relationships with such organizations?
8. Are you giving fish or teaching people to fish? Write your reflections; it is challenging to fish without technology, isn't it?
9. How would you adjust the personnel management diagram to the administration of your leaders, employees, volunteers, and consultants or contractors?
10. How can you apply the definition of technology to the function of helping others?
11. Create a SWOT analysis diagram of the technology for assistance in your ministry.
12. Make a detailed list of the people, systems, processes, policies, and assistance devices following the TAIL model applied to your ministry.
13. Create a visual map by categories of all the assistance devices you currently have or would like to have in your ministry. Use this map to create assertive plans to extend assistance in your ministry.

Chapter 10: Technology to Associate

1. Can you accompany others with technology? How?
2. How does technology strengthen communication to enhance relationships?
3. What is *koinonía pneumatos*? How does it apply to your ministry?
4. How is your ministry *koinosyncratic* and not idiosyncratic?
5. According to the concept of *koinonía*, what is the relationship between accompaniment and administration?

6. How can you use technology to accompany others in their faith journey?
7. Adapt the circles of accompaniment to your own context and create your own circles of people dedicated to accompanying others.
8. Group your processes for associating with technology into two categories: processes for sharing faith in community and processes for managing the faith community.
9. Are you using technology to manage your community? Why? What do you still need?
10. How can you apply the definition of technology to the function of accompanying others in their faith journey?
11. Create a SWOT analysis diagram of the technology for accompaniment in your ministry.
12. Make a detailed list of the people, systems, processes, policies, and accompanying devices following the TAIL model applied to your ministry.
13. Create a visual map by categories of all the accompanying devices you currently have or would like to have in your ministry. Use this map to create assertive plans to extend accompaniment in your ministry.

Chapter 11: An App for Everything

Use the chapter reference, identify the technology area you need, and visit the website links to see the details.

If you would like to obtain an electronic copy of the discussion questionnaire and all other assessment tools for printing, visit www.cyberministry.info.

ACKNOWLEDGMENTS

To God, because I owe everything to Him. He is the ultimate technologist and manager, having created the most extraordinary eternal plans and endowed human beings with unique creative and collaborative abilities within their species. The Chief Cyberminister.

I want to thank the other cyberministers in my life, who have accompanied me in my walk of faith and, perhaps unknowingly, have shaped my thinking about the church and its mission and the role of the Bible as an agent of culture and worship.

Adoring: To Mario and Tamayz Aman, for directing my gaze to the author of life worthy of all our love and dedication. To Danilo Montero for bringing me to the feet of the Master so many times, and to many other worshipers who have shaped and expressed my heart through their songs of praise and worship.

Assimilating: To Antonio Madriz, for the quiet years in the living room of his house that allowed me to see revelation with different eyes and grow as a disciple of Christ. To my seminary professors, especially Dr. Carlos Van Engen, for teaching me to think with the Bible in a holistic way; Dr. Daniel Shaw, for giving me the tools to read culture; and Dr. Pablo Deiros, for rediscovering for me the value of the history of Christianity and its relevance.

Announcing: To my cousin, Alcides Sanchez, who recruited me to paste those posters for Yiye Avila's campaign until the early morning hours and was the first to "send me out" to evangelize as my mentor in the Evangelism Explosion International program.

Assisting: To my wife, an exceptional giver! She has given me the best of herself, her best years, and is always moving to help others. To Dr. Marcel Ponton, my friend for many years, mentor, and older brother, for teaching me the importance of connecting multiple resources to help those in need, and to do it well with wisdom; his generosity inspires me.

Associating: To my pastor, Samuel Olson. For always being there when I needed him to bring clarity to my path at the most important crossroads in my life. For being a visionary man of mission. For being the first, before any literature I know of and cited in this book, to show me how the five essential functions of the mission are put into practice in managing ministry. For showing me what a life lived in the Spirit could be like. For imagining how transforming a city can look like.

To Juan Carlos Rodriguez, for welcoming me to his technical school and opening a door that has led me to over 25 years of an IT career and to contribute to the ministry through this reflection.

Finally, to all those who in one way or another helped in formulating ideas, writing, editing, formatting, publishing, and promoting this material, thank you.

CONSULTING SERVICES

There are currently several recognized Christian consultants who assist churches and other organizations in various areas of ministry. However, the availability of Christian consultants trained as theologians and technologists at the same time is limited.

The need to address the technological demands of 21st-century organizations requires what I call "bridge people" who have a good understanding of the role of technology but can develop technology plans with theological sensitivity.

That is why I have developed a model of computer consulting services known as the *Ministry Technologies Matrix®* to assist you and your ministry in three fundamental areas:

- **Coach**: Associated with training staff in the technical tasks of using and applying various technologies, providing training for both technical and ministry personnel.

- **Construct**: Associated with establishing a strong digital presence through the implementation of websites, email, social media, and mobile applications.

- **Consult**: Associated with implementing the *Ministry Technologies Matrix®* within your ministry by creating a *Custom Maturity Roadmap*™ that applies the concepts of this book to guide you towards fulfilling your mission in context.

I invite you to contact me to explore each of these three service areas I offer or the certification program. I can customize my work for short-term, medium-term, or long-term engagements, depending on your current needs.

You can use my online contact page for more information:

https://vladimirlugo.com/contact/

ABOUT THE SERIES

The topic of technology is extensive. That is why we need to explore the use of technology in church ministry from various angles.

Each volume in this series provides a theoretical framework, born from theological and managerial reflection on technology, while also offering practical advice based on technological and ministerial experience.

In other words, each book will show you the dock and how to swim to stay safe in these times of turbulent change.

Cyberministry 1: *Cyberministry: 5 Essential Functions of Technology in the Mission of the Church*

Cyberministry 2: *Cybermaturity 7 Steps to Unlock the Power of Technology for Fruitful Ministry*

Also available in Spanish as *Ciberministerio* and *Cibermadurez*.

Visit this website if you want to stay informed about this series and other books in the collection:

www.cyberministry.info

ABOUT THE AUTHOR

You devise the commission
before you commission the device.

– VLADIMIR LUGO.

Vladimir Lugo is a ministerial technology advisor, IT professional, writer, speaker, and minister by calling, who assists Christian organizations in adopting online ministerial technologies to unleash the power of their mission.

As an IT professional, he has over twenty years of experience. He has developed a deep understanding of information systems and has helped many institutions align their use of technology with their respective missions and strategic goals.

As a minister by calling, he has served the Christian community for over thirty years in ministry, understanding his calling as an educator and mentor. He has early adopted computerized programs for Bible study, social media, blogs, and electronic applications for teaching and learning. He has assisted various ministries in implementing administrative, productivity, and online presence systems. He studied Chemical Engineering in his home country, Venezuela, and graduated with a Master of Divinity (M.Div.) from Fuller Theological Seminary in Pasadena, California, with a concentration in Intercultural Studies. He holds specialist certifications in various Oracle technologies and as an *Oracle Cloud* and *Amazon Web Services* Certified Solutions Architect.

He worked for 13 years at Loyola Marymount University, where he managed the software infrastructure for all administrative systems, which is a technical way of saying that if there was a database and a web server, he was responsible for it.

He worked for Ellucian, a company that provides software solutions for close to three thousand universities and colleges, migrating these systems to the public cloud before forming his own IT consulting company, Tepui Cloud, and content development company, Tepui Media, for which he currently works. He spends his time reading because he believes that readers deserve the best. You can follow his reading habits on the social reading platform *Goodreads*:

https://goodreads.com/vladimirlugomt

He dedicates time to writing about leadership and the use of technology in Christian ministry through his blog, www.vladimirlugo.com. There, you can share your email to receive his articles: Cyberminister today, where he explores the topic of ministerial technologies week after week.

Currently, he is preparing other books for publication. By sharing your email, you will also stay informed about his future projects and opportunities to hear him speak live.

Together with his wife of over thirty years, Ivonne, they are parents to two daughters, Johanna, and Natasha. They live in the Silicon Beach area of California. He enjoys running, cycling, and swimming and frequently participates in triathlons and marathons. He loves God, loves his women, loves the Church, and loves the adventure of life.

You can check his detailed work and service history on his LinkedIn profile:

https://linkedin.com/in/vladimirlugomt

He is active on social media with the unique handle *@vladimirlugomt*. Join the conversation on social media about the book using the hashtag #cyberministry and about ministerial technologies using the hashtag #ministerialtechnologies.

If you would like to invite him to speak at your next event:

https://vladimirlugo.com/contract/conference/

Reach him out via electronic mail at: vladimir@vladimirlugo.com or by scanning the following QR code:

LIMITATION OF LIABILITY

The author authored this book to provide information. He has made every effort to ensure that the information is accurate and comprehensive, but there may be typographical or content errors.

The purpose of this book is to educate. The author and the publisher do not guarantee that the information contained in this book is complete and shall not be liable for any errors or omissions.

The author and the publisher limit their legal liability for damages caused or alleged to be caused, directly or indirectly, by the content of this book.

If you have any claims or legal questions, please contact the author at: legal@vladimirlugo.com.

Limitation of Liability

BIBLIOGRAPHY

Books

Aristotle, *Aristotle's Physics*, Hippocrates G. Apostle, trans., The Peripatetic Press (Grinnell, IA: 1980).

Aristotle, *Ars Rhetorica*, W. D. Ross, ed., Clarendon Press (Medford, MA: 1959).

Aristotle, *Nicomachean Ethics*, Second ed., Terence Irwin, trans., Hackett Publishing Company (Indianapolis, IN: 1999).

Arndt, W., Danker, F. W., y Bauer, W., *A Greek-English lexicon of the New Testament and other early Christian literature*, 3rd ed., University of Chicago Press (Chicago, IL: 2000).

Brannan, Rick, *Historic Creeds and Confessions*, Lexham Press (Bellingham, WA: 2001).

Comunidades Europeas, *Lenguaje y tecnología: De la torre de Babel a la aldea global*, European Commission, ed., (Brussels, Belgium: 1997).

Coogan, Michael D., ed., *New Oxford Annotated Bible with the Apocrypha: New Revised Standard Version.*, 3rd. Edition, Oxford University Press (New York: NY, 2001).

Covey, Stephen R., *The 7 Habits of Highly Effective People*, Mango Publishing (Coral Gables, FL: 2003).

Craig, Angela L., *Online Jesus: A Guide to Building Community, Disciple & Care Online*, Angela Craig (Seattle, WA: 2020).

Cremin, Lawrence A., *Traditions of American Education*, Basic Books (New York, NY: 1977).

Deiros, Pablo A., *El Espíritu Santo Hoy*, Serie: Programa de formación ministerial por extensión, Publicaciones Proforme (Buenos Aires, Argentina: 2000).

Fisher, Neal F., *The Parables of Jesus: Glimpses of God's Reign*, The Crossroad Publishing Company (New York: NY, 1990).

Freeman, James M. y Harold J. Chadwick, *Manners & Customs of the Bible,* Bridge-Logos Publishers (North Brunswick, NJ: 1998).

Garrett, Jeremiah K., "Sons of the Prophets," *The Lexham Bible Dictionary*, John D. Barry et al. Ed., Lexham Press (Bellingham, WA: 2016).

Gebel, Dante, *Destinado al Éxito: Cómo Alcanzar tus Mejores Sueños*, Zondervan, (Nashville, TN: 2011).

Gomez, David, *Evangelizar en Internet: Descubre cómo usar Internet para proclamar el evangelio hasta lo último de la tierra*, David Gómez (San Sebastián, Spain: 2017).

Heidegger, Martin, *The Question Concerning Technology and Other Essays*, Harper Perennial Modern Thoughts (New York, NY: 2013).

Hiebert, Paul G., *Cultural Anthropology*, Second Edition, Seventh Impression, Baker Book House, (Grand Rapids, MI: 1995).

Hirsh, Alan, *5Q: Reactivating the Original Intelligence and Capacity of the Body of Christ*, 100M (Cody, WY: 2017).

Illich, Ivan, *In the Vineyard of the Text: A Commentary to Hugh's Didascalicon*, University of Chicago Press (Chicago, IL: 1996).

Isaacson, Walter, *Invent & Wander: The Collected Writings of Jeff Bezos*, Harvard Business Review (Boston, MA: 2021).

Johnson, Clay A, *The Information Diet: A Case for Conscious Consumption*, 1st ed., O'Reilly Media, (Sebastopol, CA: 2012).

Kelly, Kevin, *What Technology Wants,* Viking (New York, NY: 2010).

Klingberg, Torkel, *The Overflowing Brain: Information Overload and the Limits of Working Memory*, Oxford University Press (New York, NY: 2009).

Liddell, H. G., *A Lexicon: Abridged from Liddell and Scott's Greek-English Lexicon*, Logos Research Systems, Inc. (Oak Harbor, WA: 1996).

Lugo, Vladimir, *Cybermaturity, 7 Steps to Unlock the Power of Technology for Fruitful Ministry*, Tepui Media (Sheridan, WY: 2023).

Madison, Dan, *Process Mapping, Process Improvement and Process Management: A Practical Guide to Enhancing Work and Information Flow*, Paton Professional Press, (Chico, CA: 2005).

Malphurs, Aubrey, *A New Kind of Church: Understanding Models of Ministry in the 21st Century*, Baker Books (Grand Rapids, MI: 2007).

__, *Advanced Strategic Planning: A 21st Century Model for Church and Ministry Leaders*, Baker Books (Grand Rapids, MI: 2013).

__ and Gordon E. Penfold, *re:Vision: The Key to Transforming Your Church*, Baker Books (Grand Rapids, MI: 2007).

Mancini, Will, *Church Unique: How Missional Leaders Cast Vision, Capture Culture, and Create Movement*, Jossey-Bass (San Francisco, CA: 2008).

Mathews, Kenneth A., "The Historical Books," in *Holman Concise Bible Commentary*, ed. David S. Dockery, Broadman & Holman Publishers (Nashville, TN: 1998).

Mitcham, Carl, *Thinking through Technology: The Path between Engineering and Philosophy*, University of Chicago Press (Chicago, IL: 1994).

Myers, Edward, "Hugh of Saint-Victor" *The Catholic Encyclopedia: An International Work of Reference on the Constitution, Doctrine, Discipline, and History of the Catholic Church*, Vol. VII, Herbermann, Charles G., Edward A. Pace, Condé B. Pallen, et. al., ed, The Encyclopedia Press (New York, NY: 1913).

Niebuhr, H. Richard, *Christ and Culture*, Harper & Row Publishers (New York, NY: 1956).

Pazmiño, Robert W., *Foundational Issues in Christian Education: An Introduction in Evangelical Perspective*, 2nd Ed., Baker Books (Grand Rapids: MI, 2001).

Proust, Marcel, *The Captive*, C. K. Scott Moncrieff trans., Global Gray (London, England: 2021).

Roberts, Alexander and James Donaldson, eds., *Ante-Nicene Fathers, Volume I: The Apostolic Fathers with Justin Martyr and Irenaeus*, Charles Scribner's Sons (New York, NY, 1903).

Rosenthal, Mark and Pavel Yudin, ed., *A Dictionary of Philosophy*, Progress Publishers (Moscow, Russia: 1967).

Sale, Kirkpatrick. *After Eden: The Evolution of Human Domination*, Duke University Press, (Durham, NC: 2006).

Sauvajol-Rialland, Caroline, *InfobéSité: Comprendre Et Maîtriser La Déferlante D'informations Qui Nos Submerge,* (Paris, France: 2013).

Spadaro, Antonio, *Cybertheology: Thinking Christianity in the Era of the Internet*, Maria Way, trans., Fordham University Press (New York, NY: 2014).

Stott, John R. W., *The Message of Acts to the Ends of the Earth*, The Bible Speaks Today, InterVarsity Press (Downers Grove, IL: 1990).

Summers, Harry G., *American Strategy in Vietnam: A Critical Analysis*, Dover Publications (Mineola, NY: 2007).

Taylor, Jerome (tr. ed.), *The Didascalicon of Hugh, of Saint-Victor: A Medieval Guide to the Arts*, Columbia University Press, Reprint edition (New York, NY: 1991).

Tesla, Nikola, *Collier's Interview with Nikola Tesla*, by John. B Kennedy, Collier's The National Weekly, vol. 27, no. 5 (Springfield, OH: 1926).

Tzu Sun, *The Art of War*, Lionel Giles trans., William Collins (London, England: 2019).

Utley, Bob, *Luke the Historian: The Book of Acts*, Study Guide Commentary Series, Bible Lessons International (Marshall, TX: 2015).

Van Engen, Charles, *God's Missionary People: Rethinking the Purpose of the Local Church*, Baker Academic (Grand Rapids, MI: 1991).

Waltner, James H., "Psalms," *Believers Church Bible Commentary*, Herald Press (Waterloo, ON: 2006).

Warren, Rick, *The Purpose-Driven Church*, Zondervan (Grand Rapids, MI: 1995).

Websites

Bible Gateway, HarperCollins Christian Publishing, [https://www.biblegateway.com], last accessed on June 15, 2023.

Digital Project Manager, *How to Create A RACI Chart: What Project Managers Need to Know*, 2023 [https://thedigitalprojectmanager.com/projects/leadership-team-management/raci-chart-made-simple/], last accessed on June 15, 2023.

Holmes, Ryan, *5 Free Apps to Simplify your Digital Life*, Inc. Magazine, (New York, NY: 2013) [https://www.inc.com/ryan-holmes/5-free-apps-to-simplify-your-digital-life.html], last accessed on June 15, 2023.

Introduction to Web Accessibility, World Wide Web Consortium W3C (Cambridge, MA: 2019) [https://www.w3.org/WAI/fundamentals/accessibility-intro/], last accessed on June 15, 2023.

IT History Society, *Casio PB-100* (Tiburon, CA: 2019), [https://www.ithistory.org/db/hardware/casio-computer-co-ltd/casio-pb-100], last accessed on June 15, 2023.

Key, Allan, *Wikiquotes*, [http://en.wikiquote.org/wiki/Alan_Kay], last accessed on June 15, 2023.

Lugo, Vladimir, *3 Passages to Formulate a Biblical Theology of Technology* (Los Angeles, CA: 2015) [https://vladimirlugo.com/3-passages-to-formulate-a-biblical-theology-of-technology/], last accessed on June 15, 2023.

__, *5 Advantages you will Enjoy with a Clear Business Model* (Los Angeles, CA: 2018) [https://vladimirlugo. com/5-advantages-you-will-enjoy-with-a-clear-business-model/], last accessed on June 15, 2023.

__, *5 Columns that Sustain the Use of Technology in the Church*, (Los Angeles, CA: 2016) [https://vladimirlugo.com/5-columns-that-sustain-the-use-of-technology-in-the-church/], last accessed on June 15, 2023.

__, *5 Components your Strategic Plan Cannot Leave Out*, (Los Angeles, CA: 2015) [https://vladimirlugo.com/5-components-your-strategic-plan-cannot-leave-out/], last accessed on June 15, 2023.

__, *5 Evils Inherent to Technology we are Called to Heal* (Los Angeles, CA: 2017) [https://vladimirlugo.com/5-evils-inherent-to-technology-we-are-called-to-heal/], last accessed on June 15, 2023.

__, *5 Lessons Today's Church can Learn from Social Networks* (Los Angeles, CA: 2015) [https://vladimirlugo.com/5-lessons-todays-church-can-learn-from-social-networks/], last accessed on June 15, 2023.

__, *7 Sources of Free Photos to Use in your Visual Presence* (Los Angeles, CA: 2017) [https://vladimirlugo.com/7-sources-of-free-photos-to-use-in-your-visual-presence/], last accessed on June 15, 2023.

Merriam Webster Dictionary, 2011 [https://www. merriam-webster.com], last accessed on June 15, 2023.

Mosendz, Polly, *Apple Launches Investigation into Celebrity iCloud Hack*, The Wire, [https://www.theatlantic.com/technology/archive/2014/09/apple-launches-investigation-into-celebrity-icloud-hack/379439/], last accessed on June 15, 2023.

Oxford Dictionaries Online quarterly update: new words added to oxforddictionaries.com today, Oxford Dictionaries, 2013, [https://public.oed.com/updates/new-words-list-june-2013/], last accessed on June 15, 2023.

SANS Institute, *Information Security Policy Templates*, 2015, [http://www.sans.org/security-resources/policies/], last accessed on June 15, 2023.

Sierra, Elena Jorge, *Actitudes Ante la Ciencia y la Tecnología: 'Apocalípticos e Integrados.' Tecnorealismo y Determinismo*

Tecnológico [https://sociotecno4.wordpress.com/actitudes-ante-la-ciencia-y-la-tecnologia-apocalipticos-e-integrados/], last accessed on June 15, 2023.

Stones, Alison, *Scriptorium: the term and its history,* Open Edition Journals [https://journals.openedition.org/perspective/4401], last accessed on June 15, 2023

The Didaché, Early Christian Writings [https://www.earlychristianwritings.com/text/didache-roberts.html], last accessed on June 15, 2023.

Vargas, Daniel R., *¿Qué dijo Billy Graham desde Puerto Rico?* Primera Hora, [https://www.primerahora.com/noticias/puerto-rico/notas/que-dijo-billy-graham-desde-puerto-rico/], last accessed on June 15, 2023.

ENDNOTES

All translations into English by Vladimir Lugo.

1 For a deeper study of Hugh of Saint-Victor's influential, please review Edward Myers, "Hugh of Saint-Victor" *The Catholic Encyclopedia: An International Work of Reference on the Constitution, Doctrine, Discipline, and History of the Catholic Church*, Vol. VII, Herbermann, Charles G., Edward A. Pace, Condé B. Pallen, et. al., ed, The Encyclopedia Press (New York, NY: 1913), pp. 521-523; Taylor, Jerome (tr. ed.), *The Didascalicon of Hugh, of Saint-Victor: A Medieval Guide to the Arts*, Columbia University Press; Reprint edition (New York, NY: 1991); and Illich, Ivan, *In the Vineyard of the Text: A Commentary to Hugh's Didascalicon*, University of Chicago Press (Chicago, IL: 1996).

2 Taylor, *Op. Cit.*, French textual references translated by the author.

3 Aristotle, *Aristotle's Physics*, Hippocrates G. Apostle, trans., The Peripatetic Press (Grinnell, IA: 1980), p. 28.

4 IT History Society, *Casio PB-100* (Tiburon, CA: 2019), [https://www.ithistory.org/db/hardware/casio-computer-co-ltd/casio-pb-100], last accessed on June 15, 2023.

5 Proust, Marcel, *The Captive*, C. K. Scott Moncrieff trans., Global Gray (London, England: 2021), p. 148.

6 Merriam Webster Dictionary, *ubiquity,* 2011 [https://www.merriam-webster.com], accessed for the last time June 15, 2023.

7 Kelly, Kevin, *What Technology Wants,* Kindle Edition, Viking (New York, NY: 2010), loc. 2621.

8 *Ibid.*, loc. 2574.

[9] Lugo, Vladimir, *5 Evils Inherent to Technology we are Called to Heal* (Los Angeles, CA: 2017) [https://vladimirlugo.com/5-evils-inherent-to-technology-we-are-called-to-heal/], last accessed on June 15, 2023.

[10] Sale, Kirkpatrick, *After Eden: The Evolution of Human Domination*, Duke University Press (Durham, NC: 2006). This book deals extensively with ludism, a movement that opposed the industrial revolution in England with ripple effects well into the end of the Twenty-First Century as neoludism. Ludists are technophobes who oppose artificial intelligence, nuclear energy, and information technology, among other forms of technology.

[11] Sierra, Elena Jorge, *Actitudes Ante la Ciencia y la Tecnología: 'Apocalípticos e Integrados.' Tecnorealismo y Determinismo Tecnológico* [http://sociotecno4.wordpress.com/actitudes-ante-la-ciencia-y-la-tecnologia-apocalipticos-e-integrados/], last accessed on June 15, 2023.

[12] Kelly, Kevin. *Óp. cit.*, loc. 618.

[13] Niebuhr, H. Richard, *Christ and Culture*, Harper & Row Publishers (New York, NY: 1956).

[14] See, for example, Kevin Kelly, *Op. Cit.*, who explains how the Amish have developed a selection criterion for using modern technologies and even develop technologies of their own.

[15] Numbers 26. The census of Israel before entering the land reveals more than 600,000 men over 20 years old capable of going to war. If we add women, elders, and children, it could easily estimate a population of around three million people.

[16] Merriam Webster Dictionary, *discipline,* 2011 [https://www.merriam-webster.com], last accessed on June 15, 2023.

[17] Aristotle, *Nicomachean Ethics*, Second ed., Terence Irwin, trans., Hackett Publishing Company (Indianapolis, IN: 1999), p. 10.

[18] This word got first introduced in the book *Infobesity: Understanding and Controlling the Information Flow that Overwhelms Us,* from the French original text not yet available in English. Caroline Sauvajol-Rialland, *InfobéSité: Comprendre Et Maîtriser La Déferlante D'informations Qui Nos Submerge*, Vuibert (Paris, France: 2013).

[19] Much has been written about this topic at the managerial, organizational, and even scientific levels. Dr. Torkel Klingberg, a professor of Cognitive Neuroscience at the Karolinska Institute

in Stockholm, writes that in our capacity to process information "boundaries are not defined by technology but by our own biology". Torkel Klingberg, *The Overflowing Brain: Information Overload and the Limits of Working Memory*, Oxford University Press, (New York, NY: 2009), p. 3.

[20] This is a neologism coined by the author, derived from another book published in English: Clay A. Johnson, *The Information Diet: A Case for Conscious Consumption*, 1st ed., O'Reilly Media, (Sebastopol, CA: 2012).

[21] Mosendz, Polly, *Apple Launches Investigation into Celebrity iCloud Hack*, The Wire, [https://www.theatlantic.com/technology/archive/2014/09/apple-launches-investigation-into-celebrity-icloud-hack/379439/], last accessed on June 15, 2023.

[22] Open source refers to the availability of the original source code of the software, while free software refers to applications that are free of charge. Many applications fall into both categories or some combination of these.

[23] Philippians 2:13: "for it is God who works in you to will and to act in order to fulfill his good purpose."

[24] Kelly, Kevin, *Op. Cit.*, loc. 2869.

[25] With co-author John P. Holdren, in *Saturday Review* (June 5, 1971), 54, 59. Ehrlich is making his own paraphrase of an earlier quote, citing Aldo Leopold.

[26] According to Oxford Dictionary, we add more than one thousand words every year. *Oxford Dictionaries Online quarterly update: new words added to oxforddictionaries.com today*, Oxford Dictionaries, 2013 [https://public.oed.com/updates/new-words-list-june-2013], last accessed on June 15, 2023.

[27] Key, Allan, *Wikiquotes* [http://en.wikiquote.org/wiki/Alan_Kay], last accessed on June 15, 2023.

[28] Merriam Webster Dictionary, *technology,* 2011 [https://www.merriam-webster.com], accessed for the last time June 15, 2023.

[29] Aristotle, *Ars Rhetorica*, W. D. Ross, ed., Clarendon Press (Medford, MA: 1959), pp. 1354b, 1355a, 1356a. Technology translates from the Greek as 'systematizing or establishing the art rules' in this case of rhetoric as art), following H. G. Liddell, *A Lexicon: Abridged from Liddell and Scott's Greek-English Lexicon*, Logos Research Systems, Inc. (Oak Harbor, WA: 1996), p. 804.

[30] Mitcham, Carl, *Thinking through Technology: The Path between Engineering and Philosophy*, University of Chicago Press (Chicago, IL: 1994), pp. 128-131.

[31] Kelly, Kevin, *Op. cit.*, loc. 159.

[32] Hiebert, Paul G., *Cultural Anthropology*, Second Edition, Seventh Impression, Baker Book House (Grand Rapids, MI: 1995), p. 62.

[33] Lugo, Vladimir, *Cybermaturity, 7 Steps to Unlock the Power of Technology for Fruitful Ministry*, Tepui Media (Sheridan, WY: 2023), Chapter 10.

[34] Merriam Webster Dictionary, *system,* 2011 [https://www.merriam-webster.com], last accessed on June 15, 2023.

[35] Return on Investment (ROI) and Total Cost of Ownership (TCO) are in fact standard calculated values many businesses use in their sales approach. Ask for these whenever you evaluate the purchase of any equipment or software application.

[36] Madison, Dan, *Process Mapping, Process Improvement and Process Management: A Practical Guide to Enhancing Work and Information Flow*, Paton Professional Press (Chico, CA: 2005), p. 3.

[37] Gebel, Dante, *Destinado al Éxito: Cómo Alcanzar tus Mejores Sueños*, Zondervan, Edición Kindle (Nashville, TN: 2011), loc. 2090.

[38] *Ibid.*, loc. 2114.

[39] SANS Institute, *Information Security Policy Templates*, 2015, [http://www.sans.org/security-resources/policies/], last accessed on June 15, 2023.

[40] You may consult the Internet for more information about the RACI matrix, but I recommend that you start with this article by the Digital Project Manager, *How to Create A RACI Chart: What Project Managers Need to Know*, 2023 [https://thedigitalprojectmanager.com/projects/leadership-team-management/raci-chart-made-simple/], last accessed on June 15, 2023.

[41] Tzu Sun, *The Art of War*, Lionel Giles trans., William Collins (London, England: 2019), 2:19.

[42] Arndt, W., Danker, F. W., y Bauer, W., *A Greek-English lexicon of the New Testament and other early Christian literature*, 3rd ed., University of Chicago Press (Chicago, IL: 2000), p.948.

[43] Roberts, Alexander and James Donaldson, eds., *Ante-Nicene Fathers, Volume I: The Apostolic Fathers with Justin Martyr and Irenaeus*, Charles Scribner's Sons (New York, NY, 1903), p. 15.

[44] Malphurs, Aubrey, *Advanced Strategic Planning: A 21st-Centuy Model for Church and Ministry Leaders*, Baker Books (Grand Rapids, MI: 2013).

[45] Mancini, Will, *Church Unique: How Missional Leaders Cast Vision, Capture Culture, and Create Movement*, Jossey-Bass (San Francisco, CA: 2008), pp. 70-76.

[46] Malphurs, Aubrey and Gordon E. Penfold, *re:Vision: The Key to Transforming Your Church*, Baker Books (Grand Rapids, MI: 2014), pp. 168-169.

[47] Review the Models section in Chapter 11 for a definition of each of these methodologies. I also wrote an introductory article to Business Model Generation titled *5 Advantages you will Enjoy with a Clear Business Model* (Los Angeles, CA: 2018) [https://vladimirlugo.com/5-advantages-you-will-enjoy-with-a-clear-business-model/], last accessed on June 15, 2023.

[48] Roberts, *Op. cit.*

[49] Summers, Harry G., *American Strategy in Vietnam: A Critical Analysis*, Dover Publications (Mineola, NY: 2007), p. 115.

[50] Merriam Webster Dictionary, *tactics*, 2011 [https://www.merriam-webster.com], accessed for the last time June 15, 2023.

[51] See, for example, Heraclitus of Ephesus, in *A Dictionary of Philosophy*, Progress Publishers (Moscow, Russia: 1967), p. 188, edited by Mark Rosenthal and Pavel Yudin. According to Heraclitus, everything in nature is in constant change, following certain rules that prevent change from descending into complete chaos. These rules that maintain controlled change without deviating from their purposes are called Logos. Heraclitus was the first philosopher to attribute metaphysical status to Logos, and he certainly influenced the introduction of the Gospel of John, where Logos is the principle of constant change.

[52] Isaacson, Walter, *Invent & Wander: The Collected Writings of Jeff Bezos*, Harvard Business Review (Boston, MA: 2021), p. 83.

[53] Lugo, Vladimir, *3 Passages to Formulate a Biblical Theology of Technology* (Los Angeles, CA: 2015) [https://vladimirlugo.com/3-passages-to-formulate-a-biblical-theology-of-technology/], last accessed on June 15, 2023.

[54] You can read more details about *kybernéseis* in the second book of the series: *Cybermaturity,* Op. cit., Chapter 8.

[55] Van Engen, Carlos, *God's Missionary People: Rethinking the Purpose of the Local Church*, Baker Academic (Grand Rapids, MI: 1991).

[56] Warren, Rick, *The Purpose-Driven Church*, Zondervan (Grand Rapids, MI: 1995). pp. 121-126.

[57] Malphurs, Aubrey, *A New Kind of Church: Understanding Models of Ministry for the 21st Century*, Baker Books (Grand Rapids, MI: 2007), P. 102.

[58] Hirsh, Alan, *5Q: Reactivating the Original Intelligence and Capacity of the Body of Christ*, 100M (Cody, WY: 2017), p. 64.

[59] *Ibid.*, p. 66.

[60] The metaverse is a new concept that became the talk of the media when, in 2021, Facebook decided to change its name to Meta. The metaverse is a conglomerate of new technologies that includes online gaming, video conferencing, avatars and digital worlds, digital ownership through non-fungible tokens (NFTs), and cryptocurrencies. The church today is called to exercise its functions in these new digital worlds, hence the new concept of the metaversal church.

[61] Review page 174 for an explanation of phylacteries.

[62] Lugo, Vladimir, *5 Lessons Today's Church can Learn from Social Networks* (Los Angeles, CA: 2015) [https://vladimirlugo.com/5-lessons-todays-church-can-learn-from-social-networks/], last accessed on June 15, 2023.

[63] Mathews, Kenneth A., "The Historical Books," in *Holman Concise Bible Commentary*, ed. David S. Dockery (Nashville, TN: Broadman & Holman Publishers, 1998), p. 161.

[64] Waltner, James H., "Psalms," *Believers Church Bible Commentary* (Scottdale, PA; Waterloo, ON: Herald Press, 2006), p. 759.

[65] Cremin, Lawrence A., *Traditions of American Education*, (New York, NY: Basic Books, 1977), p. 134.

[66] Comunidades Europeas, *Lenguaje y tecnología: De la torre de Babel a la aldea global*, Comisión Europea, ed., (Luxembourg: Brussels, 1997), p. 1.

[67] *The Didaché's* entire text is publicly available at *Early Christian Writings*, https://www.earlychristianwritings.com/text/didache-roberts.html, last accessed on June 15, 2023.

[68] Stones, Alison, *Scriptorium: the term and its history,* Open Edition Journals [https://journals.openedition.org/perspective/4401], last accessed on June 15, 2023.

[69] Fisher, Neal F., *The Parables of Jesus: Glimpses of God's Reign*, The Crossroad Publishing Company (New York: NY, 1990), pp. 21-22.

[70] Pazmiño, Robert W., *Foundational Issues in Christian Education: An Introduction in Evangelical Perspective*, 2nd Ed., Baker Books (Grand Rapids: MI, 2001), p. 45.

[71] See for example James M. Freeman y Harold J. Chadwick, in *Manners & Customs of the Bible*, Bridge-Logos Publishers (North Brunswick, NJ: 1998), p. 256; or Jeremiah K. Garrett, "Sons of the Prophets," ed. John D. Barry et al., *The Lexham Bible Dictionary*, Lexham Press (Bellingham, WA: 2016).

[72] Accreditation of your educational program can include national or regional agencies, the endorsement of ecclesiastical denominations, affiliation with independent organizations such as Educause, the Association for Biblical Higher Education (ABHE), or the Transnational Association of Christian Colleges and Schools (TRACS), or participation in institutional accreditors backed by the Department of Education, such as the Association of Theological Schools (ATS), among others.

[73] Vargas, Daniel R., *¿Qué dijo Billy Graham desde Puerto Rico?* Primera Hora, [https://www.primerahora.com/noticias/puerto-rico/notas/que-dijo-billy-graham-desde-puerto-rico/], last accessed on June 15, 2023.

[74] Warren, Rick, *Op. Cit.*, p. 154.

[75] GDPR stands for General Data Protection Regulation in Europe, and CCPA stands for California Consumer Privacy Act, which is the equivalent regulation to protect consumer privacy.

[76] Cookies are small text files that are stored on a device when visiting a website. These files contain information that the website can use to remember certain data about the user, such as their preferences or login information. Cookies are used by websites to provide a more personalized and efficient experience to the user, as they allow for information to be remembered and activity on the website to be tracked.

[77] Lugo, Vladimir, *7 Sources of Free Photos to Use in your Visual Presence* (Los Angeles, CA: 2017) [https://vladimirlugo.com/7-sources-of-free-photos-to-use-in-your-visual-presence/], last accessed on June 15, 2023.

[78] __, *Cybermaturity*, Op. Cit., Chapter 5.

[79] Coogan, Michael D., ed., *New Oxford Annotated Bible with the Apocrypha: New Revised Standard Version.*, 3rd. Edition, Oxford University Press (New York: NY, 2001), Sirach 50:25–26. Added parenthesis from the Septuagint. The Greek *ethne* could be equally translated as people or nation, and from which we derive the English word "ethnic."

[80] Covey, Stephen R., *The 7 Habits of Highly Effective People*, Mango Publishing (Coral Gables, FL: 2003), p. 163. Second habit.

[81] Utley, Bob, *Luke the Historian: The Book of Acts*, Study Guide Commentary Series, Bible Lessons International (Marshall, TX: 2015), Acts 6.

[82] *Introduction to Web Accessibility*, World Wide Web Consortium W3C (Cambridge, MA: 2019) [https://www.w3.org/WAI/fundamentals/accessibility-intro/], last accessed on June 15, 2023.

[83] Stott, John R. W., *The Message of Acts to the Ends of the Earth*, The Bible Speaks Today (Downers Grove, IL: InterVarsity Press, 1990), p. 105.

[84] Gomez, David, *Evangelizar en Internet: Descubre cómo usar Internet para proclamar el evangelio hasta lo último de la tierra*, David Gómez (San Sebastián, Spain: 2017), p. 9.

[85] Brannan, Rick, *Historic Creeds and Confessions*, Lexham Press (Bellingham, WA: 2001), Nicene Creed.

[86] Deiros, Pablo A., *El Espíritu Santo Hoy*, Serie: Programa de formación ministerial por extensión, Publicaciones Proforme (Buenos Aires, Argentina: 2000), p. 131.

[87] Craig, Angela L., *Online Jesus: A Guide to Building Community, Disciple & Care Online*, Angela Craig (Seattle, WA: 2020), p. 20.

[88] Tesla, Nikola, *Collier's Interview with Nikola Tesla*, by John. B Kennedy, Collier's The National Weekly, vol. 27, no. 5, January 30, 1926 (Springfield, OH: 1926), p. 16.

[89] Holmes, Ryan, *5 Free Apps to Simplify your Digital Life*, Inc. Magazine, (New York, NY: 2013) [https://www.inc.com/ryan-holmes/5-free-apps-to-simplify-your-digital-life.html], last accessed on June 15, 2023.

This book has been published by Tepui Media.

Tepui Media is a company committed to strengthening the initiatives of independent authors and amplifying their communication impact. Tepui Media has a team of highly qualified professionals in multimedia communications, including specialists in digital presence and marketing, audiovisual production, content strategy, and software development.

If you need to publish your book and create a strong platform for promoting and disseminating your message, talk to the experts:

https://tepui.media

info@tepui.media

Vladimir Lugo

Made in the United States
Orlando, FL
[illegible]

Made in United States
Orlando, FL
04 November 2026

90200514R00197